SpringerBriefs in Materials

More information about this series at http://www.springer.com/series/10111

Les Vickers · Arie van Riessen
William D.A. Rickard

Fire-Resistant Geopolymers

Role of Fibres and Fillers to Enhance
Thermal Properties

Les Vickers
Geopolymer Research Group
Curtin University
Perth, WA
Australia

William D.A. Rickard
Geopolymer Research Group
Curtin University
Perth, WA
Australia

Arie van Riessen
Geopolymer Research Group
Curtin University
Perth, WA
Australia

ISSN 2192-1091 ISSN 2192-1105 (electronic)
SpringerBriefs in Materials
ISBN 978-981-287-310-1 ISBN 978-981-287-311-8 (eBook)
DOI 10.1007/978-981-287-311-8

Library of Congress Control Number: 2014955620

Springer Singapore Heidelberg New York Dordrecht London
© The Author(s) 2015
This work is subject to copyright. All rights are reserved by the Publisher, whether the whole or part of the material is concerned, specifically the rights of translation, reprinting, reuse of illustrations, recitation, broadcasting, reproduction on microfilms or in any other physical way, and transmission or information storage and retrieval, electronic adaptation, computer software, or by similar or dissimilar methodology now known or hereafter developed.
The use of general descriptive names, registered names, trademarks, service marks, etc. in this publication does not imply, even in the absence of a specific statement, that such names are exempt from the relevant protective laws and regulations and therefore free for general use.
The publisher, the authors and the editors are safe to assume that the advice and information in this book are believed to be true and accurate at the date of publication. Neither the publisher nor the authors or the editors give a warranty, express or implied, with respect to the material contained herein or for any errors or omissions that may have been made.

Printed on acid-free paper

Springer Science+Business Media Singapore Pte Ltd. is part of Springer Science+Business Media (www.springer.com)

Preface

This review is designed to introduce the reader to some of the basic concepts of geopolymer (or alkali activated materials) science and technology with the aim of demonstrating the applicability of geopolymers as fire-resistant products. Although geopolymers have impressive fire-resistant properties, improvements can be made by adding fibres and/or fillers, and a considerable portion of this book is dedicated to this aspect of enhancing the thermal properties of these materials.

An important feature of geopolymers and their composite products is that they offer a real and superior alternative to products based on Ordinary Portland Cement (OPC). As such we have provided extensive information about OPC to enable comparisons to be made with geopolymer-based products.

This text is intended to fill a gap that exists between short and very specifically focussed research papers and lengthy textbooks that endeavour to cover all aspects of geopolymerisation. The contents of this book are targeted at those committed to promoting use of new materials with a low carbon footprint.

Contents

1 **Introduction to Geopolymers** 1
   1.1 Overview of Geopolymers 1
   1.2 History of Geopolymers 3
   1.3 Portland Cement (OPC) and Concrete 4
   1.4 Geopolymer Applications 10

2 **Precursors and Additives for Geopolymer Synthesis** 17
   2.1 Alkali Soluble Aluminosilicate Sources 18
   2.2 Alkaline Dissolution Media 23
   2.3 Admixtures and Fillers for Geopolymer Systems 29
       2.3.1 Admixtures 29
       2.3.2 Fillers 34

3 **Chemistry of Geopolymers** 39
   3.1 Metakaolin Based Geopolymers 40
   3.2 Fly Ash Based Geopolymers 47
   3.3 The Role of Calcium Compounds 51

4 **Fibres: Technical Benefits** 53
   4.1 Reinforcement 56
   4.2 Steel Fibre Reinforced Concrete (SFRC) 57
   4.3 Organic Fibres 60
       4.3.1 Polypropylene and Other Polyolefin Fibres 64
       4.3.2 Polyvinyl Alcohol Fibres 66
       4.3.3 Other Organic Fibres 67
       4.3.4 Carbon Based Reinforcing Fibres 68
   4.4 Inorganic Fibres 72

5	**Thermal Properties of Geopolymers**		77
	5.1	Measurement of Thermal Transport Properties	78
	5.2	Thermal Expansion	81
		5.2.1 Thermal Expansion of Geopolymers	83
	5.3	Thermal Conductivity	94
		5.3.1 Thermal Conductivity of Geopolymers	95
6	**Fire Resistance of OPC and Geopolymers**		99
	6.1	Fire Testing	100
	6.2	Portland Cement	103
	6.3	Fire Resistance of Geopolymers	105
7	**Conclusions**		111
	References		113

Chapter 1
Introduction to Geopolymers

1.1 Overview of Geopolymers

Geopolymers, also referred to as Aluminosilicate Inorganic Polymers (AIP) and Alkali Activated Cement (AAC) are based on alkali soluble aluminium and silicon precursors (aluminosilicates). Structural differences and resulting properties of geopolymers can be explained by variation in the source silicon to aluminium amorphous molar ratio, alkali metal cation type and concentration, water content and curing regime amongst other variables in the geopolymer synthesis.

Figure 1.1 shows geopolymers to be part of the alkali activated family of cementitious materials, characterised by low calcium content. Early work by Purdon (1940) using sodium hydroxide solutions to activate ground blast furnace slags (GBFS) produced cementitous materials suitable for concrete production. Later workers showed that these materials were basically a calcium silicate hydrate based gel (Roy 1999; Wang and Scrivener 1995) with silicon present mainly in one dimensional chains and some substitution of aluminium for silicon and magnesium for calcium whereas the geopolymer gel is a three dimensional alkali aluminosilicate framework structure (Duxson et al. 2007). Calcium (~3 wt%) in geopolymers acts as an accelerator for setting enabling ambient temperature curing of geopolymers to take place (Temuujin et al. 2009b). The defining characteristic of a geopolymer is that the binding phase consists of an alkali alumina silicate gel, with aluminium and silicon linked in a three dimensional tetrahedral gel framework of silicate and aluminate groups that is relatively resistant to dissolution in water. Charge balancing of the aluminate group is by alkali metal cations typically sodium and potassium.

Geopolymer may be considered a composite of partially reacted precursor, typically metakaolin or fly ash, solid reaction products, and pore space. Some of the pore space is filled with aqueous alkaline solution. The pore volume can be in the order of 1–40 %. The solid reaction product is amorphous and is the inorganic polymer component (Maitland et al. 2011).

Geopolymers are inorganic materials based on the polymerisation of silicon and aluminium tetrahedra precursors in highly alkaline media with the alkali or

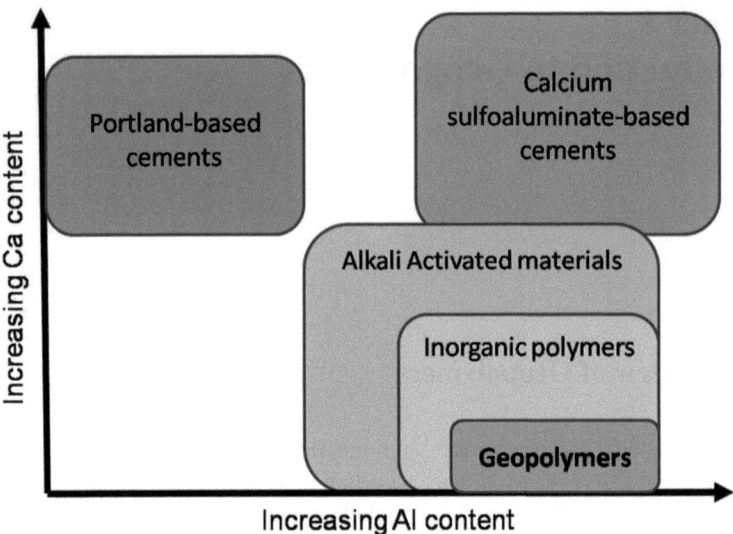

Fig. 1.1 Schematic showing the chemical relationship of geopolymers to Portland and other cements. Adapted from Provis (2014)

alkaline earth metal cations providing charge balancing to the Al(IV) co-ordinated anion (Barbosa et al. 2000a). They are essentially amorphous in structure and can act as cementitious binders for mortars and concretes in the construction industry (Hardjito and Rangan 2005), where they exhibit improved durability when compared to products based on Portland cement binders (Roy 1999). This durability exhibits itself as superior chemical and heat, including fire, resistance. They have also been extensively researched as binders to immobilise toxic heavy metals e.g. from spent nuclear fuel and other toxic waste (van Jaarsveld et al. 1997) and as fabric laminating resins in the manufacture of high temperature composites (Lyon 1999).

The impetus for the utilisation of geopolymers is their ability to consume readily available industrial by-products in their synthesis with the potential to reduce by-product stock piles, reduce dependency on shrinking natural resources and give a net reduction in carbon dioxide emissions from cement manufacture. World cement production in 2011 was estimated at 3.4 billion tonnes (van Oss 2012), which equates to carbon dioxide emissions attributed to clinker of 2.57 billion tonnes (Mehta and Meryman 2009). Many of these industrial by-products are being stock piled in tailings dams which will contribute to environmental issues in the event of weather extremes such as high rain fall. These events can lead to overflowing and/or ruptured retaining walls with resultant flooding and water supply contamination. A case in point is the Kingston Fossil plant in Roane County, Tennessee in December 2008 when a period of twice average rainfall and freeze thaw cycles broke the retaining wall and deposited fly ash sludge over 12 ha and up to 1.8 m deep before spilling into the Emory river (Stephens 2009).

The supply of suitable industrial by-products, particularly fly ash and blast furnace slag, is limited by location (place of production) and quantity. They are already being used as Supplementary Cementitious Materials (SCMs) to reduce the proportion of Portland cement used in concrete manufacture, which in turn reduces the available quantity for AAC binders. In the Third World production of fly ash and blast furnace slag is limited so alternative SCMs such as clays need to be developed for local cementitious binder production (Scrivener 2011). The use of volcanic ashes is also a potentially useful raw material for AAC production (Lemougna et al. 2011).

Geopolymer precursors can be based on dehydroxylated clay i.e. metakaolin, and a wide range of industrial and agricultural by-products such as fly ash, aluminium and steel slags, spent metal treatment solutions (Nugteren et al. 2011), fumed silicas from ceramic raw material preparation, rice husk ashes (Bernal et al. 2012b), and crushed glass (Naik 2002). The precursors need to have significant amounts of aluminium and/or silicon entities in amorphous form to facilitate the alkali solubilisation and resulting geopolymer formation.

1.2 History of Geopolymers

Ancient mortars and concrete products have proven to be more durable and more resistant to acid attack and freeze-thaw cycles than products based on Ordinary Portland Cement (OPC) as shown by rapid failure of OPC repair materials on ancient concretes (Pacheco-Torgal et al. 2008a; Roy 1999). The intrinsic properties of OPC based product leads to high permeability that enables water and other aggressive media to enter and take part in detrimental chemical reactions, particularly with the ever present calcium hydroxide (Pacheco-Torgal et al. 2008b). The Greeks and Romans used a concrete to construct buildings such as the Pantheon and the Coliseum. These concretes were based on cement derived from lime and pozzolans such as volcanic ash and clays. One theory for building the Cheops Pyramid at Gaza is the use of formed in-situ blocks based on alkali activated aluminosilicate materials (Demortier 2004). Historically cementitious materials were being used in Sumeria (3600 BC) (Bauer 2007), Egypt (2500 BC) (Davidovits 2008c), Rome (Davidovits and Davidovits 1999; Davidovits 2008a).

In 1940 Purdon carried out work on alkali activation of blast furnace slag. This work was based on 31 Belgian slags and used sodium hydroxide solutions of up to 10 wt% concentration with maximum compressive strength (25 MPa for concrete) achieved in the range of 5–8 wt% (pH values of 14.1–14.3). He also investigated the replacement of sodium hydroxide with blends of calcium hydroxide and sodium carbonate which react to produce sodium hydroxide in the presence of water. He showed that the sodium hydroxide liberated by the latter reaction brings about the setting of the slag. The addition of calcium hydroxide and sodium carbonate to the slag produced cement requiring only the addition of water to bring about setting. He also found that the alkali reactivity of the slag increased as the chilling temperature of the slag decreased with a corresponding increase in the amorphous content. Purdon proposed that the alkali acted as a catalyst for slag setting and hardening. This was confirmed by recovering all the added alkali from a set slag mortar. This explains the lower sodium hydroxide levels used here compared to geopolymer formation. He also identified one of the present drawbacks of geopolymer concrete manufacture—the use of corrosive alkaline solutions for the dissolution process (Purdon 1940).

Glukhovsky and subsequently Krivenko in the USSR (Ukraine) from the 1950s onwards developed alkali activated binder systems due to the shortage of OPC raw materials in the then state controlled economy. Glukhovsky was the first to appreciate that the natural processes of the transformation of volcanic rocks into zeolites could be copied and carried out in cementitious systems to yield useful construction products. Glukhovsky analysed binders used by the Romans and Ancient Egyptians and based on this work developed binders called "soil cements". These were based on aluminosilicate containing slags mixed with alkali industrial waste. In the 1960s applications such as housing, railway sleepers, drainage and irrigation channels, flooring, and precast slabs and blocks were produced using these slag based systems (Krivenko 2005).

Davidovits began investigating alkali activated cements following catastrophic fires in France in 1970 and 1973. The goal was to develop heat and fire resistant materials in the form of non-flammable "plastic like" materials. Davidovits gave the name geopolymers to these new materials which were predominantly based on metakaolin (Davidovits 1989).

From the mid 1990s work by Palomo's group in Spain (1999), Rahier's (1997) work in Belgium and the group at the University of Melbourne (Xu and van Deventer 2000; van Jaarsveld and van Deventer 1999) began to de-mystify the geopolymer reaction kinetics and mechanisms. In the United States groups led by Balaguru (Lyon et al. 1997) and Kriven (2008) explored potential geopolymer applications in the fire resistant composite and refractory areas, respectively. MacKenzie (Nicholson et al. 2005) in New Zealand has focussed on the synthesis and analysis, particularly NMR, of geopolymers. Sanjayan (Kong et al. 2005) at Monash University and van Riessen et al. (2010) at Curtin University currently lead research teams carrying out geopolymer and precursor analysis and thermal property investigations with a strong emphasis on fire testing.

Geopolymer research is now wide spread both academically and industrially with greater emphasis being placed on application based research as commercialisation gathers momentum.

1.3 Portland Cement (OPC) and Concrete

The forerunner of Portland cement was developed in the late 18th century, when Smeaton calcined limestone and clay to form a cementitious material. Joseph Aspdin took out the patent for Portland cement in 1824. The Portland name was adopted because the colour of the hydrated cement was similar to that of limestone

1.3 Portland Cement (OPC) and Concrete

quarried at Portland in southern England. Over the next 200 years it became the major construction material (Shi and Mo 2008).

Portland cement initially gave variable properties due to the low calcination temperatures used to prepare the clinker which was finely ground and used as cement. In 1845 Isaac Johnson first burnt the raw materials in a glass making kiln, rather than a cooler lime kiln, at the required clinkering temperature (1,400–1,500 °C) to produce cement resembling the current era material (Gani 1997).

The main application of OPC is to make concrete. Plain concrete made from OPC and aggregate is referred to as first generation concrete. Second generation concrete refers to steel bar reinforcement invented by Joseph Monier in 1849. Coignet patented a technique in 1856 for reinforcing concrete using iron tirants. The first reinforced concrete bridge was built in 1889 in the Golden Gate Park in California (Li 2011).

Prestressed concrete is referred to as third generation concrete. This was developed in the 1880s to overcome cracking in reinforced concrete. Prestressing is usually generated by the stretched reinforcing steel in a structural member. Prestressed concrete became accepted as a building material in Europe after WWII due to the shortage of steel. The Walnut Lane Memorial Bridge in Philadelphia, completed in 1951 was the first prestressed concrete structure in North America.

Compressive strength at an age of 28 days is the main design index for concrete due to concrete's use in structures mainly to resist compressive forces. The measurement of compressive strength is relatively easy and it is thought that other properties can be related to compressive strength. The pursuit of high compressive strength has been an important goal in concrete development.

Duff Adams in 1918 found that compressive strength of concrete is inversely proportional to the water-cement ratio (w/c). Hence a high compressive strength is obtained by reducing the w/c ratio, but there is a minimum water requirement to maintain workability. For this reason progress in developing high compressive strength was slow prior to the 1960s. At that time 30 MPa compressive strength development was considered to be high strength concrete (Li 2011).

Since the 1960s the development of high compressive strength has progressed due to two main factors, the invention of water reducing admixtures and the incorporation of mineral admixtures such as slag, fly ash and silica fume.

Water reducing admixtures help maintain good workability at low w/c ratios. The mineral admixtures have a small particle size and can react with calcium hydroxide, a hydration product in concrete, to give a denser microstructure. Silica fume also has a packing effect to further increase matrix density.

Concrete produced after the 1980s usually contains mineral and chemical admixtures so that the hydration mechanism and resulting hydration products and microstructure are very different to concrete produced without their addition. These new generation concretes are referred to as contemporary concrete (Li 2011).

Two developments of note with contemporary concrete are self-compacting concrete (SCC) and high performance concrete (HPC). Concretes with compressive strengths of around 130 MPa are termed ultra-high performance concrete and are used in the latest generation of skyscrapers (Li 2011).

SCC meets special requirements which cannot be achieved using conventional materials and techniques. The requirements may involve enhancement of installation techniques involving placement and compaction without segregation plus improved durability. SCC (with high flowability obtained with the addition of superplasticisers) is used to fill formwork, typically with closely spaced rebar, without the need for mechanical vibration. SCC was developed in Japan where highly reinforced concrete is required for earthquake resistance.

It became clear that many concrete structures could not fulfil service demands due to a lack of durability. The development of HPC overcame some of these durability issues, but created specific issues when heat as in fire exposure became involved.

OPC is manufactured by milling raw materials such as limestone, clay and shale with an iron source. This blend is now fired in a rotary kiln at 1,400–1,500 °C. The resulting clinker is then cooled, ground again with gypsum, which acts as a set retarder, to give 90–95 % of material finer than 325 mesh (10–15 μm median particle size). The alkali metal content is kept below 1 % to minimise alkali silica reactions with added aggregates. Cement chemistry will vary from area to area because of variations in local raw materials. Typical ASTM C150 Portland cements are made up from 50 to 70 wt% tricalcium silicate, (C_3S), 15 to 20 wt% dicalcium silicate, (C_2S), 5 to 10 wt% tricalcium aluminate, (C_3A), 5 to 15 wt% ferrite phase and 3 to 5 wt% gypsum (Vitro Minerals 2010).

The presence of iron in conjunction with aluminium in the feedstock has a marked effect on liquid formation in the kiln, reducing the temperature from 2,065 °C to the usual operating range of 1,400–1,500 °C (MacLaren and White 2003).

Best quality cement requires the presence of tricalcium silicate (C_3S) and dicalcium silicate (C_2S) in the clinker (MacLaren and White 2003). This paper also gives a comprehensive listing of the shorthand notations used to simplify descriptions of cement compositions. Both these materials react vigorously with water to produce the cement binder paste. Tricalcium silicate reacts and sets much faster than dicalcium silicate (hours vs. days).

When cement hydrates the principle products formed are 50–70 wt% calcium silicate hydrate (CSH), 10–15 wt% ettringite (calcium sulfoaluminate) and 20–25 wt% calcium hydroxide (CH). CSH is the strength building binder for concrete but the CH makes no contribution to strength and can lead to efflorescence and poor chemical resistance. Replacing cement with a pozzolan reduces the formation of CH (by dilution) whilst the pozzolan reacts with the remaining CH to form additional CSH binder with improvements in properties (Vitro Minerals 2010).

All of the compounds capable of hardening do so at different rates and generate different quantities of heat per unit weight, but only silicates contribute to strength (Central Federal Lands Highway Division 2008).

The typical heat generation stages due to cement hydration are shown in Fig. 1.2. During the initial hydration on mixing with water calcium and hydroxyl ions are released from the C_3S surface causing the pH level to rise. When calcium and hydroxyl values reach a critical level, crystallisation of calcium hydroxide and CSH begins. These initial reactions are temperature dependent. This is followed

1.3 Portland Cement (OPC) and Concrete

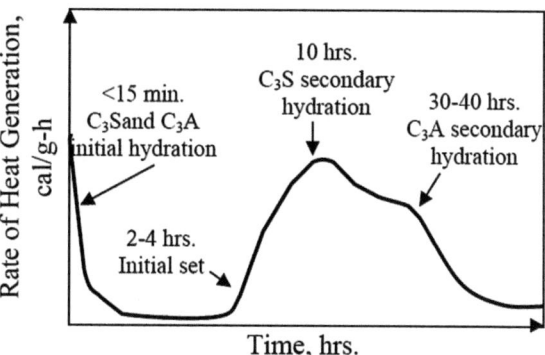

Fig. 1.2 Typical rate of heat evolution during setting of cement (Central Federal Lands Highway Division 2008)

by a dormant stage (no/little heat evolved) where the cement remains plastic. Calcium hydroxide crystallises from the solution and CSH develops on the surface of C_3S forming a coating. As the thickness of this coating increases the rate of water penetration decreases and the speed of the reaction becomes diffusion controlled. The C_2S is hydrating at a slower rate due to its lower reactivity. The end of this period is the initial set and acceleration of the hydration now occurs as the critical concentration of ions is reached and silicates hydrate rapidly. Maximum hydration rate occurs within this time span. The rate of reaction now slows and is completely diffusion dependent (Vitro Minerals 2010).

The rate and quantity of heat evolution is a function of cement chemical composition, cement fineness and particle size distribution, w/c ratio and reaction temperature. The reaction temperature is variable due to the heat of hydration and the size (surface area to volume ratio) of the cast part. Temperature increases of up to 40 °C are possible (Central Federal Lands Highway Division 2008). The increase in temperature can affect the microstructure of the concrete reducing the molecular size of phases with resultant weakening and greater propensity to cracking.

Large structures such as dams, tunnel linings and drilled shafts can generate significant quantities of heat leading to temperature differentials between the interior volume and exterior surface of the concrete. Temperature gradients can also arise in sections of varying thicknesses. Measures are required to manage these thermal effects such as internal cooling and exterior insulation. Maximum allowable temperature differentials are usually specified. In the United States this difference is commonly set to 20 °C (Central Federal Lands Highway Division 2008).

Heat generated during concrete curing leads to thermal stresses and subsequent concrete cracking. Controls to combat thermal issues include using low heat generating cement and the use of aggregates with low coefficients of thermal expansion (COTE). These measures help to reduce the effects of heat build-up and can contribute to the performance and durability of the finished product. Reducing the placement temperature of the concrete helps reduce the rate of hydration with a corresponding reduction in peak temperature (Neville 1995).

The quantity of cement present in a mix is the significant factor in heat generation. This is true whether the amount of cement is reduced, the water increased or pozzolans added. There are limits to these modifications as they also affect the intrinsic properties of the final concrete. All these measures reduce cracking and rigidity and the compressive and tensile strengths.

The action of water is a common factor in the processes of concrete degradation. Hardened cement is porous, containing a dual network of pores. The capillary pore system is characterised by pores of 50–1,000 nm diameter and extends throughout the system acting as channels between the various components/phases. The cement gel contains a network of gel pores of 10–50 nm diameter. Physical properties of cement systems such as stiffness, fire resistance and durability are directly related to the amount of water present. Hardened cement generally contains 30–40 % water occurring in 3 forms (MacLaren and White 2003):

- Chemically bound water i.e. water of hydration which is chemically bound to cement precursors in the form of hydrates. This accounts for approximately 90 % of the water.
- Physically bound water. This is water adsorbed onto the surface of the capillaries. Mainly found in the small gel pores of the system.
- Free water is found within the larger pores and is able to flow in and out of the system. The amount depends on the pore structure and volume, relative humidity and presence of water in direct contact with the hardened cement surface such as in water bearing pipes and marine structures.

The ability of water to dissolve cement components and its volume changes in freeze-thaw cycles can cause durability issues. Free water is particularly significant in durability issues as it is able move throughout the hardened cement.

As cement paste hydrates over several months the porosity decreases. Initially the drying process is by capillary flow of water through the larger pores. As the porosity decreases the water transport process becomes diffusive in nature. Higher w/c ratios result in larger pore sizes as the cement gel forms and these contain a larger volume of water. Larger pores give a faster drying rate which can create problems in low relative humidity curing conditions. If cement is exposed to long periods of low humidity and high temperatures, adsorbed water in the gel pores of the cement will evaporate. This process leads to drying shrinkage. Partially filled gel pores contain water menisci which exert tensile stresses on the walls of the pores leading to micro-cracking.

Freeze-thaw cycles occur when temperatures hover around 0 °C and can occur on a daily basis. Water shows a 9 % volume increase on freezing. This expansion exerts stresses on the capillary walls causing micro-cracking. These micro-cracks fill up with water during the thaw cycle, freeze when the temperature drops causing more cracks to open (Neville 1995).

Crystalline calcium hydroxide makes up about 10 vol.% in commonly used systems. Calcium hydroxide has an ambient temperature solubility in water of around 1.7 g/l and is readily dissolved in the free water in pores (MacLaren and White 2003). Removal of calcium hydroxide by dissolution leaves voids in the system which can encourage deeper penetration of water. Carbon dioxide from the atmosphere can react with this calcium hydroxide in solution. The visible signs of this process, known as efflorescence, are the formation of white deposits of

1.3 Portland Cement (OPC) and Concrete

calcium carbonate on the exterior surfaces of the concrete. A similar reaction with calcium silicate hydrate is also possible (Neville 1995).

Water can also carry aggressive chemicals into the cement paste where they attack the various components. Acidic attack reduces the pH of the pore water which will promote corrosion of steel reinforcement members. The intrusion of chloride ions from sea water and de-icing compounds accelerates corrosive attack on steel reinforcement. Corrosive salts, particularly ammonium and magnesium sulphates, react with calcium hydroxide to form calcium sulphate which has approximately twice the volume of the removed calcium hydroxide. This increase in volume sets up internal stresses leading to cracking (MacLaren and White 2003).

The alkali silica reaction (ASR) is caused by silica compounds in certain aggregates reacting with alkalis from within the concrete or from external sources e.g. de-icing salts, ground water and sea water. The result of the reaction is an expansive process leading to longitudinal, map or pattern cracking, spalls at joints and overall deterioration. The chemical reaction between soluble silica in the aggregate and soluble alkali produces an alkaline silica gel that swells when water is absorbed. The swelling of the gel may crack the concrete. Existing cracks can be filled with gel preventing them from closing and causing further cracking (Transportation Research Board 2006). The type, size and amount of reactive aggregate play important roles in the reaction.

Early methods of prevention included limiting the total alkali content of cement below 0.6 wt% of cement, but experience showed that < 0.4 wt% was required to prevent ASR occurring. When high amounts of alkali are present in cement, pozzolans (fly ash, silica fume and metakaolin), blast furnace slag, or lithium salts are used to inhibit the ASR reaction. Pozzolans and slag are effective because they tie up hydroxyl ions thereby preventing formation of expansive gel. They also reduce alkali concentrations by replacing portions of the OPC and reduce permeability which prevents penetration of alkalis from external sources. Lithium salts are thought to function by the formation of a non-swelling gel. Lowering internal relative humidity to less than 80 % also stops ASR expansion (Ferraris 1995).

The presence of ettringite, calcium sulphoaluminate, found in all OPC concrete can lead to long term crack formation. Gypsum and other sulphate compounds react with calcium aluminate in the cement to form ettringite within the first few hours after mixing with water. Essentially all the sulphur in the cement is consumed to form ettringite within 24 h. The formation of ettringite results in a volume increase in fresh, plastic concrete. However due to the plastic nature of the concrete this expansion is harmless (Portland Cement Association 2012).

In the longer term ettringite will dissolve in water and reform in less confined areas e.g. white needle like crystals can be seen lining air voids. The phenomenon of delayed ettringite formation (DEF) is usually associated with heat cured concrete. The high temperature, typically 70 °C, decomposes any initially formed ettringite and the resulting sulphate and alumina entities are tightly bound in the CSH gel. In the presence of moisture the sulphate and alumina desorb from the CSH to form ettringite in confined locations in cooled, hardened concrete. Since the paste is rigid and there are insufficient voids to accommodate the ettringite volume increase expansion and resultant cracking occurs. In addition some of the initially formed ettringite may be converted to monosulphoaluminate at higher temperatures and on cooling revert back to ettringite. Ettringite takes up more space than the monosulphoaluminate from which it forms and the transition is therefore an expansive process (Collepardi 2000).

In the case of fire and thermal resistance the transport of water through the cement body is essential to reduce spalling tendencies. The existing pore system can be supplemented by the addition of small diameter thermoplastic fibres which melt as the temperature increases forming channels for water transport. In the case of HPC, the addition of silica fume gives a dense microstructure with reduced pore interconnectivity. Thermoplastic fibres are essential in these systems to increase the interconnectivity of the pores, which facilitates water transport (Papworth 2000; Kalifa et al. 2001).

The mechanical properties and volume stability characteristics of OPC concrete are markedly reduced by exposure to elevated temperatures, which can lead to structural failures (Arioz 2007). Release of chemically bound water from CSH becomes significant above 110 °C. This dehydration and the thermal expansion of any aggregates increase internal stresses and beyond 300 °C micro-cracking occurs in the system. Calcium hydroxide dissociates at 530 °C resulting in shrinkage. Water used to fight the fire results in the regeneration of calcium hydroxide resulting in cracking and crumbling of the concrete. Most changes occurring above 500 °C are considered irreversible. CSH gel undergoes further dehydration above 600 °C with the formation of cracks.

The interfacial transition zone (ITZ) between cement paste and aggregate is described as the most important interface in concrete (Scrivener et al. 2004). It has a major impact on strength and permeability. The ITZ arises from the "wall" effect of packing of cement grains against a relatively flat aggregate surface. The zone closest to the aggregate contains predominantly small grains and significantly higher porosity with coarser grains further out. In concrete this means that the size of the ITZ is comparable to cement particles (10–50 µm). The ITZ is generally weaker than either paste or aggregate due to high local w/c ratio and particle packing issues. In some cases large crystals of calcium hydroxide and ettringite are orientated perpendicular to the aggregate surface. Micro-cracking is common in the ITZ which can result in shear bond failure and interconnected macro-porosity, which influences permeability. The most effective way to modify the ITZ is to add a fine particle material such as 5–10 % fume silica. This is common practise in the production of high strength concrete HPC (Scrivener et al. 2004).

1.4 Geopolymer Applications

The hardened geopolymer has an amorphous glass-like structure which is capable of modification by aggregates, reinforcing agents and process aids during the mixing and shaping process. Potential applications include

1.4 Geopolymer Applications

concretes and mortars (in competition with OPC) (Hardjito and Rangan 2005; Skvara et al. 2006), specialised high temperature applications such as fire and heat resistant products (with superior durability to OPC), binders for encapsulation of toxic chemicals and nuclear waste and chemically resistant products (Rostami and Brendley 2003; Nugteren et al. 2011; Palomo and Fernandez-Jimenez 2011).

In the short term the use of geopolymer systems for OPC replacement is likely to be limited to special high performance applications where their outstanding properties compared to traditional materials gives them a clear technical advantage. These limitations are based on a lack of long term performance data, a shortage of geopolymer feedstock compared to total OPC consumed and higher current costs together with adverse OH&S scenarios with Class 8 corrosive liquid alkaline activators (Scrivener 2011).

Typical high performance applications are shown in Table 1.1. They highlight the three key property areas, chemical, heat and fire resistance, where geopolymers have demonstrated a clear superiority and structural stability over OPC based products.

Table 1.1 Potential high performance applications for geopolymers

Military
(i) Exhaust gas thermal insulation: VTOL landing pads, Carrier launch deflector plates, armoured vehicles. Naval turbine engine exhaust
(ii) Marine: Fire resistant components
(iii) Ammunition storage bunkers
(iv) Fire resistant fuel storage bunds
Civil
(i) Chemical resistant bunding and flooring e.g. sewerage treatment, dairy floors, electrowinning cells
(ii) Solar power heat transfer components
(iii) Oil and Gas down hole cementing
(iv) Heat resistant coatings for steel, wood and existing OPC structures
(v) Tunnel linings
(vi) Bench tops
(vii) Extruded profiles for fire resistant door jambs and window frames
(viii) Elevator components, for passive fire protection in high rise buildings
(ix) Thermal Insulation systems including cellular products
(x) Railway sleepers
(xi) Sewerage pipes
Mining
(i) Prefabricated components (chemical and fire resistant) for power generation and mineral processing) e.g. ducting
(ii) Tunnel linings
(iii) Geothermal grouting
(iv) Back fill
Hazardous waste management
(i) Nuclear waste storage
(ii) Toxic metal encapsulation
(iii) Water purification

The United States military have been a major driver of new technologies and the resultant higher service conditions placed on construction materials. The introduction of the AV8B (Harrier) vertical takeoff and landing (VTOL) aircraft led to spalling of concrete landing pads and degradation of asphalt based pads. The AV8B has an exhaust temperature of approximately 700 °C, which equates to a ground temperature of 450 °C. The new F35B has an exhaust temperature 205 °C higher and a ground temperature of at least twice that of the AV8B (Sweetman 2011). US Navy specifications call for vertical landing pads to be made from high temperature concrete (Ceratech 2011).

This ability to resist high temperatures coupled with inherent fire resistance will open up other military markets such as carrier launch deflector plates, armoured vehicle and marine turbine engine exhaust systems and fire walls for ammunition bunkers.

Efficient down hole cementing is essential for well recovery in low pressure and leaky strata in newer oil and gas production fields (Zhu et al. 2010). Compared to present cementing systems, the geopolymer cement investigated exhibited lower permeability, with all the pores smaller than 20 nm, and high early compressive strength (30 MPa after 24 h at 80 °C). The flowability and setting times could be modified by the use of admixtures to enable pours to be completed to large depths.

Down hole cementing in the oil and gas industry is a demanding application from an installation and service perspective. US Patent 7794537 (Barlet-Gaudedard et al. 2010) is for a geopolymer composition with controllable thickening and setting times over a wide range of temperatures by the use of accelerators and retarders. The compositions have good mixability and pumpability (viscosity in the range 200–300 mPa s). Based on metakaolin, fly ash or slag with typical a silicon to aluminium ratio of 2, the compositions exhibit strength and permeability values suitable for well cementing applications. The binder compositions may be filled with light weight fillers such as cenospheres or high density fillers such as barytes to suit the end applications.

Nasvi (2012) compared a fly ash based geopolymer to conventional well cements and found a marked reduction in CO_2 permeability. Existing well cements degrade in the acidic CO_2 environment and allow release of the CO_2 to the atmosphere via the increased porosity in the well cement.

In a similar vein, geothermal energy functions by the use of a heat pump which transfers thermal energy from the ground through a closed buried pipe system sealed with a grouting material which ensures the stability and thermal transmission of the bore hole. The grout must bond well to the pipe work and have a thermal conductivity greater than 1.0 W/m K. Cement and cement-bentonite blends have been used as binders and the fillers/aggregate thermally enhance the grout system. Whilst the use of geopolymer was not reported the use of slag as an SCM gave thermal conductivity values in the range of 1.8–2.0 W/m K (Borinago-Trevino et al. 2012). This would be an application where the superior durability of geopolymers would be advantageous.

Thermal energy storage for solar power generation is a potential market for geopolymers. The use of concrete as a storage medium along with the use of a heat transfer fluid is being evaluated (Panneer Selvam and Hale 2011). The original design involved a stainless steel tube embedded in high strength fibre reinforced concrete. The goal is to build "concrete" tubes which are durable to 600 °C with low porosity to retain the molten salt heat transfer fluid. This is an application with good potential for geopolymers.

OPC concrete is not well known for its inherent chemical resistance. This has led to the application of organic coatings based on, but not limited to, vinyl esters, epoxies and polyureas (Stavinoha 1991; Guan 2003) to reduce the ingress of corrosive liquids into OPC concrete. Polymer concrete, i.e. aggregate mixed into a polymeric binder, is also available to form monolithic chemical resistant structures. The use of organic based coatings leads to higher raw material and labour costs together with ongoing maintenance issues. There are wide ranging opportunities for chemically resistant geopolymer systems across several industry and infrastructure sectors.

Table 1.2 shows the broad spectrum of acidic media sources. All these sources are threats to cementitious binder systems in several industry and infrastructure applications. The degree of acidic aggression depends on the nature of the anions, which govern the strength of the acid, degree of dissociation and solubility of the salt formed. The aggressiveness of organic acids towards OPC increases in order of their calcium salt solubility (Zivica and Bajza 2001).

Bakharev (2005) conducted testing of fly ash based geopolymers exposed to 5 % acetic and 5 % sulphuric acids. She found that the geopolymer performed better than OPC due to the lower calcium content of the geopolymer. The composition of the alkali activators used in the synthesis of the geopolymers had a large influence on the acid resistance. The best results were obtained from geopolymer

Table 1.2 Sources of acidic media adapted from Zivica and Bajza (2001) and Oualit et al. (2012)

Industrial processes	Refineries, wood treatment and pharmaceutical production generate substituted phenols which are acidic
	Detergents can be based on sulphonic acid derivatives
	Power generation from coal emits acidic gases (SO_2, CO_2)
	Cement manufacture emits CO_2
	Food production and storage
Urban activity	Oil based transportation systems emit CO, SO_2, NO_x gases
	Wood and Oil based heating emits CO_2
	Sewage treatment generates hydrogen sulphide
Natural effects	Carbon dioxide in water
	SO_3^{-2} and SO_4^{-2} anions in peat bogs
	Soil contains huminous acids
	Sea water is a mix of inorganic and organic acids
Bacteria	Bacteria break down proteins and hydrocarbons giving methane and hydrogen sulphide
	Also reduce sulphate ions to sulphides (usually ferrous) which in turn gives hydrogen sulphide, a precursor for sulphuric acid
Water	Solvent for aggressive media
	Transport media for aggressive media and reaction products
	Constituent of reaction products in OPC e.g. ettringite

prepared using sodium hydroxide as the sole activator and curing at elevated temperature. Two months immersion in both acids gave large weight changes for OPC (>10 wt%) whilst the geopolymer showed less than 2 % weight change.

In the food storage and processing industry several organic acids, lactic, acetic and formic are commonly used or formed by bacteria attack on sucrose and animal feed meal. Whilst they are classed as weak acids they readily attack the calcium hydroxide present in OPC to produce soluble calcium salts which are leached out of the concrete reducing the pH. This results in the binding agents being left in an unstable condition. This weakened material is now easily removed by animal scuffage and cleaning (DeBelie et al. 1997).

Milk processing causes severe OPC degradation due to lactic acid attack, and again the remedy can be the use of geopolymer based systems.

OPC is the main material of construction for sewerage and waste water plant and the associated pipework. The concrete can be attacked from the inside by sulphuric acid generated by bacteria and externally by humic and carbonic acids formed naturally in the soils or from contaminated back fill (Oualit et al. 2012). This attack leads to a reduction in strength with increased risk of sewer pipe collapse. This could lead to flooding and rupture of other infrastructure services such as water and gas pipelines.

Montes and Allouche (2008) compared class C fly ash and metakaolin based geopolymers to OPC mortars for pipeline and sewage refurbishment using trenchless technology. The fly ash based geopolymer showed the best sulphuric acid resistance. Trenchless technology encompasses a wide range of techniques utilised for the installation and refurbishment of underground services with minimal surface disruption from trench excavation. The technologies include in-situ pipe replacement and relining of existing structures. Geopolymer coatings for these applications are being investigated at the Louisiana Technical University Trenchless Technology Centre (2012).

Ammonium compounds tend to act as weak acids which will attack OPC. Ammonium nitrate is widely used in explosive and fertiliser manufacture where OPC attack is widely reported. The use of geopolymer based mortars can be used for remedial work or as part of the original installation.

MacKenzie (2011) reported on several new and potential geopolymer applications. The charge balancing Na^+ cations in conventional geopolymers can be exchanged for other cations (O'Connor et al. 2010). 100 % exchange of Na^+ by K^+, NH_4^+, Ag^+ and Pb^{2+}, but less for other cations is achievable. The use of Ag^+ produces a material with strong antimicrobial properties which can be used as a bactericidal filter bed for water purification. A similar technique can be used for the removal of heavy metals from waste water streams e.g. 100 % Pb^{2+} and 72 % Cd^{2+} but this is unable to remove Hg^{2+}.

Porous cladding for passive cooling of buildings produced from metakaolin based geopolymer with the addition of fibres as pore formers have been evaluated. The fibres are removed after setting by controlled heating leaving continuous, aligned pores which can achieve up to 1 m capillary lift for water. Cladding produced in this manner can effect cooling by latent heat of evaporation of the

1.4 Geopolymer Applications

water (Okada et al. 2011). Engineering obstacles need to be overcome to give a practical system in terms of capillary lift heights.

The resistance of OPC to heat and fire is poor with rapid loss of strength above 450 °C and the manufacture of high performance concrete (HPC) exaggerates the situation due to lower (limited connectivity) porosity resulting in spalling. This spalling is a result of pressure build up caused by trapped water/steam which exceeds the tensile strength of the OPC and pieces of material are explosively ejected from the main body. The use of fibres to form pathways to allow water escape is well documented (Papworth 2000; Heo et al. 2012).

Geopolymers generally do not show this spalling behaviour. Geopolymers based on fly ash have small, well interconnected pores, which allow easier permeation of water to the surface. During heating pore volume initially increases as voids once filled with water are emptied. At higher temperatures melting of amorphous unreacted material from fly ash particles exposes additional pores from within these unreacted particles. This additional porosity increases the total interconnected porosity of the geopolymer (van Riessen et al. 2009).

Kong (2007) found that a fly ash based geopolymer had a large number of micropores (<1.25 nm). On heating up to 800 °C he believed that these micropores facilitated the escape of water from within the sample thus causing minimal damage to the specimen.

Bakharev (2006) found that in fly ash based geopolymers the cumulative pore volumes increased by 26 % after heating to 800 °C and by 29 % at 1,000 °C compared to an unheated sample. Average pore size increased from 37.6 nm (unheated) to 121 nm (800 °C) and 1,835 nm (1,000 °C).

The increased pore volume connectivity of geopolymers compared to OPC during high temperature exposure will markedly increase transport of water through the binder with associated reduced spalling.

This, together with inherent heat resistance, very low smoke and toxic gas emissions indicates that fly ash based geopolymers will show outstanding fire resistance.

The use of geopolymers for fire resistant building products is illustrated by its use for window and door surrounds replacing organic materials (Reid 2011) and fire resistant elevator doors (Krivenko 2005). The use of geopolymers imparts greater structural integrity (as shown by enhanced glass retention), lower toxic gas emissions and reduced flammability.

Similarly the replacement of High Strength Concrete (HSC) by a geopolymer based concrete is likely to give superior fire resistance due to higher mechanical properties at elevated temperatures and reduced spalling due to the geopolymer's interconnected pore system. This substitution will enhance the structural integrity of buildings and tunnel linings during and after fire events. Fire tests carried out on 75 MPa OPC and geopolymer concrete at Monash University and reported by van Riessen et al. (2009) showed the superiority of the geopolymer concrete in resistance to spalling.

The use of geopolymers reinforced by woven textiles based on carbon fibres, basalt and glass were developed for aircraft applications to overcome the shortcomings of organic fibres (flammability and smoke evolution). Lyon (Lyon 1996; Lyon et al. 1997) carried out work for the US Federal Aviation Authority and compared a range of organic polymers to a potassium geopolymer (silicon: aluminium = 32:1) in a carbon fibre composite using the following criteria: ignitability, heat release, and smoke (ASTM E-1354), flame spread index (ASTM E-162-83) and residual flexural strength (ASTM D790). The geopolymer system was found to be superior in every category.

Research on foamed geopolymers has demonstrated their viability with applications in the construction, automotive and thermal insulation industries (Liefke 1999). Applications listed by Liefke for these foams include thermal and sound insulation for plant and equipment, buildings and transportation, silencer and catalytic converter components and filtration supports.

Coatings based on geopolymers have been reported on by Temuujin et al. (2009a, 2011). Steel coupons coated with various formulations withstood thermal treatment from a gas torch without cracking. A suitable geopolymer based coating could have applications in treatment of steel, OPC and wood to enhance fire resistance and impart corrosion protection (Zhang et al. 2010). Medri (2011) used a geopolymer based on metakaolin activated with potassium silicate and potassium hydroxide to act as a binder for silicon carbide particles. These systems could be applied by brush and were cured at 80 °C. Good adhesion to substrate was observed together with low shrinkage after exposure to 1,000 °C.

Chapter 2
Precursors and Additives for Geopolymer Synthesis

Abstract The raw materials typically used in geopolymer manufacturing are described. The use of metakaolin, coal derived fly ash and slag as aluminosilicate sources is presented. The use of readily available coal fly ash in geopolymer products offers sustainability and technical benefits, especially in niche applications such as fire resistant products. There are national standards for the use of fly ash in concrete and these have some applicability to geopolymer production. For instance, beneficiation of fly ash by screening and milling can significantly improve physical properties of resultant geopolymer products. Activation of high calcium ground blast furnace slag and other slags result in more complex microstructures and are not as suitable for fire resistant products. Metakaolin is a purer source of aluminosilicate than fly ash. It is capable of ambient cures but has the penalty of high water demand which influences property development. The investigation of volcanic ashes as aluminosilicate sources has grown in importance recently. Alkaline hydroxides and silicates are used to activate the aluminosilicate sources and the sodium and potassium derivatives are commercially available. The speciation of alkali metal silicates is briefly outlined together with the influence of the alkali metal cation on this speciation. The use of sodium aluminate derivatives is examined, in particular the use of Bayer liquor, a by-product of alumina extraction from bauxite. The structure of sodium aluminate solutions is also reviewed. A final section describes the admixtures, typically superplasticisers and fibres, and fillers employed to modify the plastic and cured properties of geopolymers with emphasis on optimisation of fire resistant properties.

Keywords Fly ash · Fly ash beneficiation · Metakaolin · Sodium silicates · Sodium aluminates · Admixtures and fillers

The precursors and additives required for the synthesis of geopolymer systems are:

- Alkali soluble aluminosilicates
- Alkaline dissolution media
- Admixtures to develop specific properties
- Aggregates and fillers.

2.1 Alkali Soluble Aluminosilicate Sources

Alkali soluble aluminosilicate sources include granulated blast furnace slag (GBFS), fly ash, and volcanic ashes. Metakaolin, calcined clay, has an amorphous structure with a silicon to aluminium ratio of 1. It is used to manufacture geopolymers and can increase the aluminium content of geopolymers when used with fly ash and GBFS. Materials such as silica fume from smelting operations and from rice husk ashes do not contain aluminium, but are alkali soluble and are used to adjust silicon content of geopolymers for property control. All the above materials have utility as Supplementary Cementitious Materials (SCM) in conjunction with Portland cement.

Sources of aluminium are limited to waste anodising solutions (Nugteren et al. 2011), sodium aluminates including Bayer liquor (Jamieson et al. 2012) and aluminium smelting slag although the latter is toxic and requires additional processing prior to use. These aluminium based materials may be used to adjust aluminium levels in geopolymer synthesis. A potential source of aluminium is waste water treatment sludge which has been evaluated in brick and cement manufacture (Babatunde and Zhao 2007).

GBFS does not form geopolymers based on aluminium and silicon tetrahedral structures, but rather a calcium silicate hydrate which is also present in OPC. GBFS's calcium levels tend to be in excess of 40 wt%. GBFS can be used in blends with fly ash to give hybrid structures (Buchwald 2009; Sofi et al. 2007; Puertas et al. 2000). Alkali activated binders based on GBFS yield products with higher compressive strengths and durability compared to OPC (Fernández-Jiménez et al. 1999). The reactions of slags are dominated by small particles. Particles greater than 20 μm in size react only slowly, whilst particles less than 2 μm react completely within 24 h. To achieve these small particle sizes grinding of the granulated slag is required (Wan et al. 2004).

Fly ash is the by-product formed during power generation by the combustion of coal. It comprises fine particles which rise with the flue gases (ash that does not rise is termed bottom ash) and is collected by electrostatic precipitators or other filtration equipment prior to the flue gases reaching the exhaust chimneys.

The source and chemical composition of coal and the operating conditions of the combustion process have a major impact on the resulting fly ash chemical composition and physical properties. The main components of fly ash are silica, alumina, ferrous oxide and calcium oxide with variable amounts of carbon as shown by LOI (loss on ignition) results (Blissett and Rowson 2012). They also observed that fly ashes derived from sub-bituminous and lignite coals are characterised by higher calcium oxide and lower silica and alumina contents compared to high grade bituminous and lignite coals (see Table 2.1). Fly ash from coals containing less than 10 wt% calcium oxide consists mainly of aluminosilicate glasses and do not usually contain crystalline calcium compounds. Fly ash with greater than 15 wt% calcium oxide contained calcium aluminosilicate glasses and crystalline calcium compounds.

Table 2.1 Bulk composition of fly ash by coal type—modified from Blissett and Rowson (2012)

Component (wt%)	Bituminous	Sub-bituminous	Lignite
SiO_2	20–60	40–60	15–45
Al_2O_3	5–35	20–30	20–25
Fe_2O_3	10–40	4–10	4–15
CaO	1–12	5–30	15–40
LOI	0–15	0–3	0–5

The Ash Development Association of Australia has published The Coal Combustion Products Handbook (Gurba et al. 2007). This handbook is a useful reference source for Australian fly ashes and their current and potential applications.

The morphology of fly ash particles is controlled in the main by combustion temperature and subsequent cooling rate. SEM imaging showed a mix of solid spheres, hollow spheres (cenospheres), irregular unburnt carbon, and mineral aggregates with quartz, corundum and magnetite. Mullite and quartz are the predominant crystalline constituents of Australian fly ashes (Rickard et al. 2011; French and Smitham 2007).

Kutchko (2006) studied 12 class F fly ashes from the United States using SEM-EDS. She found mainly amorphous aluminosilicate spheres with some iron rich spheres, the latter consisting of iron oxide and amorphous aluminosilicate. The morphology of the particles was used to determine the relative amorphous and crystalline compositions. Cross sections showed the iron oxide and aluminosilicate were mixed throughout the fly ash particles. The calcium was not associated with the aluminium or silicon. In addition to the particles reported above she also found agglomerated and amorphous material, the formation being attributed to inter-particle contacts.

The amount of alkali soluble (amorphous) fly ash material is the critical quantity for geopolymer synthesis calculations (Fernandez-Jimenez 2003). Allowance must also be made for the fact that not all the amorphous material is accessible by the activating alkali solutions.

The total bulk elemental composition (as oxides) can be determined by X-ray fluorescence (XRF) and the phase abundance by quantitative X-ray diffraction (QXRD). If the composition of the crystalline component is subtracted from the total elemental composition what is left is the composition of the amorphous content some of which can react with alkaline activators to form geopolymer products. Chen-Tan et al. (2009) carried out this type of analysis on Collie fly ash. Rickard and Williams extended this work to several other Australian fly ashes with a view to determining thermal properties of the synthesised geopolymers (2010, 2011, 2012).

Font et al. designed a simplified XRD based method for determination of the glass content and mineralogy of coal based fly ashes. A glass standard based on a >99 % aluminosilicate glass was added to fly ash in known proportions and the mixtures evaluated by XRD (Font et al. 2010).

Van Jaarsveld et al. (2003) carried out XRF to get the total elemental composition, but used FTIR to determine the aluminium co-ordination. Bands at 550 and 557 cm^{-1} had previously been assigned to aluminium in octahedral co-ordination. This latter co-ordination form will not form geopolymers. He attributed differences between two Port Augusta fly ash batches to slight differences in the silica amorphous phases. It was proposed that these differences could be due to variations in the coal feed stock composition or the variation in furnace temperatures.

Recent work using dilatometry by Provis et al. (2012) claims that mechanical properties of fly ash based geopolymers can be predicted from the temperature at which the high temperature expansion peak attributed to release of strongly chemically bound water is observed. High strength samples generally display this expansion at a higher temperature than low strength samples. Samples made either with poorly reactive fly ash or with an excessive amount of silica in the activating solution also expand by up to 10–15 % below 200 °C. This was attributed to a combination of low extent of crosslinking by aluminium entities and vaporisation of added water to produce an expanded material. Additional water was added to the geopolymer synthesis mixture to give sufficient workability to fill the 4 mm diameter moulds for dilatometer test pieces. This will also reduce strength properties of these samples. This work was based on potassium silicate activating solutions for six different fly ashes. This method may have potential to screen the suitability of fly ash for geopolymer manufacture. The key outcome of this work is the influence of available (for geopolymerisation) aluminium from the fly ash on strength development.

There are national standards for fly ash for use in concrete and these also have relevance to geopolymer synthesis (ASTM Committee C09 2012a). Table 2.2 outlines the basic requirements of fly ash in ASTM C618-2 and EN450-1. EN 450-1 uses 3 LOI categories and 2 fineness categories for the initial classification of fly ash grades.

Class F fly ash is regarded as a pozzolanic material. A pozzolan has no intrinsic cementitious properties. It will react with calcium hydroxide at ambient temperatures and in the presence of water to form compounds exhibiting cementitious properties. The high calcium content of Class C fly ash will result in the formation of cementitious properties in the absence of calcium hydroxide and hence are not true pozzolans (Blissett and Rowson 2012).

Table 2.2 ASTM and EN fly ash requirements adapted from Blissett and Rowson (2012)

ASTM C618-12	$SiO_2 + Al_2O_3 + Fe_2O_3$	SO_3	Moisture	LOI	Max retained on 45 μm sieve
Class C	>50 %	<5 %	<3 %	<6 %	34 wt%
Class F	>70 %	<5 %	<3 %	<6 %	34 wt%
EN450-1	$SiO_2 + Al_2O_3 + Fe_2O_3$	SO_3	Reactive silica	LOI	Max retained on 45 μm sieve
Category A	>70 %	<3 %	>25 %	<5 %	Category N < 40 wt%
Category B	>70 %	<3 %	>25 %	2–7 %	Category S < 10 %
Category C	>70 %	<3 %	>25 %	4–9 %	

Not all of the amorphous material in fly ash reacts during alkaline dissolution. Amorphous material can be encased inside spheres and the deposit of alkali-fly ash reaction product on the spheres can act as a barrier to further dissolution. Mechanical activation of fly ash has been evaluated to increase the amount of fly ash material reacting with the alkaline solutions (van Riessen and Chen Tan 2013).

The simplest procedure is to screen the fly ash to remove particles retained on a 45 μm screen. This effectively increases the surface area of the material which has passed through the screen. Also larger sized non aluminosilicate materials such as quartz and carbon are removed. The removal of the latter is important as it has a high water demand which can result in a drop in compressive strength (Silverstrim et al. 1997).

Nugteren (2009) separated raw fly ash into six different size fractions by air classification. Geopolymers were made from these fractions in combination with GBFS using potassium silicate solution as the alkaline activator. The window of workability varied with each fly ash fraction. He found that there was no clear trend for increases in compressive strength with finer fractions. However, mix proportions were varied for each fraction and this may have impacted on the results.

Several authors have investigated mechanical activation of fly ash (Chindaprasirt and Rattanasak 2010; He et al. 2005; Kumar and Kumar 2010; Temuujin et al. 2009c; Kumar et al. 2007).

Kumar (2007) compared raw fly ash (RFA) with mechanically activated material from vibratory milling (VMFA), attrition milled (AMFA) and jet milled classified (CFA). The median particle size of the RFA of 36 μm was reduced to 5 μm (VMFA), 4.3 μm (AMFA) and 2.8 μm (CFA). FTIR showed changes in peak intensity which was attributed to the structural rearrangements occurring during milling. Geopolymers made from mechanically activated fly ashes developed compressive strengths up to 120 MPa, with VMFA giving the highest rate of strength development.

Further work by Kumar (2010) investigated the effects of extended vibratory milling times. Particle size decreased and heat flow measured during the geopolymer synthesis increased with milling time. SEM and EDS results showed a more compact microstructure and no unreacted fly ash spheres for fly ash milled for 30 min compared to unmilled fly ash. EDS results for geopolymer based on raw fly ash gave silicon to aluminium ratios in the range 2.2–2.9 and a silicon to sodium ratio in the range 0.9–1.2. Corresponding values for geopolymer based on fly ash milled for 30 min were silicon to aluminium of 2.9–3.2 and silicon to sodium of 3.5–4.2. Extended milling time also reduced setting time and increased compressive strength.

Temuujin (2009c) treated raw Collie fly ash in a vibratory mill for 60 min. SEM of the fly ash showed that the spherical morphology was substantially removed by milling. Particle size was reduced from 14.4 to 6.8 μm by the milling procedure. Initially, fast setting of the geopolymer based on milled fly ash was experienced. This was overcome by the addition of extra water which also gave suitable workability. Compressive strength, after an ambient temperature cure regime, was increased from 16 to 45 MPa by virtue of using milled fly ash. This increase in compressive strength is claimed to be due to higher rates of geopolymerisation in samples based on milled fly ash. The ability of geopolymer systems, without added calcium compounds, to cure at ambient temperatures is imparted by the use of milled fly ash.

Temuujin (2009) investigated preliminary calcination of Collie fly ash at 500 and 800 °C as a potential activation method prior to synthesising geopolymer. This preliminary calcination caused crystallisation of mullite and hematite on the fly ash surface, lowering the reactivity of the fly ash as shown by the drop in compressive strength from 55.7 MPa (uncalcined) to 44.3 MPa (calcined at 800 °C).

Metakaolin is produced by dehydroxylation of kaolin between 650 and 800 °C. At temperatures beyond this metakaolin converts to spinel, which in turn converts to mullite (McCormick 2007). Gastuche (1962) investigated the transformation of kaolin to metakaolin and concluded that (a) the reaction proceeded by stepwise removal of constitutional water from successive octahedral sheets, (b) water removal was accompanied by a gradual transformation of Al(VI) to Al(IV), (c)cohesion between individual layers is reduced and the internal surface becomes progressively more susceptible to acid attack, whereas the silica sheets still organised as a "ring structure" remain and dissolve at a somewhat slower rate.

Metakaolin is used as an SCM with OPC where it functions by reacting with calcium hydroxide to form hydrated calcium aluminates and silicoaluminates and thus improving the durability of the resultant binders. Metakaolin has a much greater surface area than cement (12,000 m^2/g vs. 333 m^2/g). This high surface area contributes to an increase in the water demand to maintain workability. The addition of superplasticisers can help to reduce this water demand (McCormick 2007).

Metakaolin was the material used by Davidovits (2008b) for the early geopolymer development and commercialisation. Since then several research groups have used metakaolin to elucidate the kinetics, mechanisms and microstructure formation of geopolymer systems (Granizo et al. 2007; Barbosa et al. 2000a; Duxon et al. 2005, 2006, 2007a, b; Duxson et al. 2006a, b, c; Fletcher et al. 2005; Rahier et al. 1997, 2007; Weng and Sagoe-Cretsil 2007; Zuhua et al. 2009b; Zhang et al. 2010; Barbosa and MacKenzie 2003a; Kamseu et al. 2010; Subaer and van Riessen 2007).

The relative purity of metakaolin compared to fly ash and GBFS makes it eminently suitable for basic research into the many aspects of geopolymers. From a practical perspective metakaolin has a high water demand which can limit strength development and the relatively high aluminium content makes it suitable as a blending component with fly ash and GBFS to reduce the silicon:aluminium ratio of geopolymer systems where required (Phair et al. 2004).

The investigation of volcanic ash for the production of geopolymers has become more common in recent years. Volcanic ash (44 wt% SiO_2, 14 wt% Al_2O_3, 10 wt% CaO and 13 wt% Fe_2O_3) from Cameroon was activated with caustic soda to synthesis geopolymer paste with a maximum compressive strength of 55 MPa after curing at 90 °C. Mortar with 40 wt% sand gave a compressive strength of 30 MPa. This level of compressive strength exceeds the requirements of ASTM C216-12, Grade SW (severe weathering). ASTM C216 specifies the requirements for face brick (solid masonry units from clay or shale). The development of building products from readily accessible volcanic ash with suitable conversion techniques could have genuine economic benefits (Lemougna et al. 2011). The second piece of work (Tchakoute Kouamoa et al. 2012) uses volcanic ash activated by alkali fusion at 550 °C. Metakaolin is used to consume the excess alkali from the fusion process and to adjust aluminium levels in the geopolymer produced by activating with sodium silicate solutions. The Romans used volcanic ash (harena fossica) from the Naples area mixed with lime to produce concrete and mortar (Davidovits and Davidovits 1999).

Other naturally occurring pozzolanic materials are suitable for geopolymer synthesis (Bondar et al. 2010). Xu and van Deventer (2000) investigated the geopolymerisation of 16 naturally occurring minerals using sodium and potassium hydroxides as activating materials and concluded that all 16 could be suitable geopolymer precursors.

2.2 Alkaline Dissolution Media

Alkaline dissolution media come from the family of alkali and alkali earth metal hydroxides, carbonates, aluminates and silicates in aqueous solution. Sodium hydroxide is the predominant material (due to cost considerations) with the other members of the Group I family used for specific applications. Calcium (Group II), as the hydroxide, is the only alkali earth used.

In general alkali hydroxides and silicates at pH values of greater than 13 are required to activate aluminosilicates for geopolymer synthesis (Khale 2007). Alkali metal carbonates as the sole activating agent are insufficiently alkaline for geopolymerisation reactions. However, sodium carbonate can react with calcium hydroxide to form sodium hydroxide in solution. Provis (2009) gives a comprehensive overview of alkali hydroxide solutions. There is a significant increase in solution viscosity of all alkali metal hydroxides at concentrations greater than 10 M. Efflorescence (formation of white crystalline sodium carbonates) is a known issue in geopolymers activated with an excess of sodium hydroxide. Atmospheric carbon dioxide reacts with the excess alkali to give white crystalline surface deposits which are mainly cosmetic in effect. Temuujin (2009b) found efflorescence containing phosphorus on ambient cured geopolymers. XRD showed this to be sodium phosphate hydrate and was assumed to derive from the 1.3 wt% P_2O_5 in the Collie fly ash.

Potassium hydroxide activation imparts onset of densification at higher temperature than in sodium activated geopolymers (Barbosa and MacKenzie 2003a; Duxon et al. 2007b). Blends of sodium and potassium hydroxides have been evaluated (Duxson et al. 2006b) with respect to thermal shrinkage of metakaolin based geopolymers with the blended cation geopolymer laying between the pure cation systems.

Alkali metal silicates, based on sodium and potassium are commercially available (PQ Europe 2004). Production can be via the hydrothermal or furnace route.

Generally silicates are identified by the $SiO_2:M_2O$ ratio, known as the modulus. Commercially this ratio is normally specified in terms of weight. To convert to a molar ratio the following multiplier factors are used, sodium 1.033 and potassium 1.567. The viscosity of sodium silicate solutions is a function of concentration, modulus and temperature. Potassium silicate solutions show similar behaviour, but are more viscous than corresponding sodium silicate solutions at equal concentration. The pH of commercial silicate solutions is in the range of 10–13. Sodium silicate solutions will polymerise to form silica gel when the pH drops below 10. Solutions of sodium silicates react with dissolved polyvalent cations such as Ca^{2+} or Al^{3+} to form insoluble forms of silicates.

Potassium silicate resembles its sodium counterpart in physical properties but also shows several advantages: it does not develop efflorescence, displays higher solubility and compatibility with other additives and imparts higher refractoriness, flowing at higher temperatures than the sodium analogues.

Sodium and potassium silicate solutions consist of a range of silicate anions (PQ Corporation 2005). The basic building block is the tetrahedral silicate anion, the monomer unit. These monomer units can link in a variety of ways to form linear, planar cyclic and three dimensional silicate anion structures. There are two main factors which influence the distribution of silicate anion types, the ratio of silica to alkali and the concentration. Figure 2.1 shows the distribution of silicate anion species as the modulus increases at constant concentration. Moving from left to right (reducing alkali content) shows a shift from high monomer ("monosilicate") to more complex structures and polymers via a series of growing chains and rings.

The distribution of anionic species can be altered by the addition of alkali or dilution with water, known as respeciation. This causes the solution to change to the species distribution appropriate to the new modulus. A full strength solution

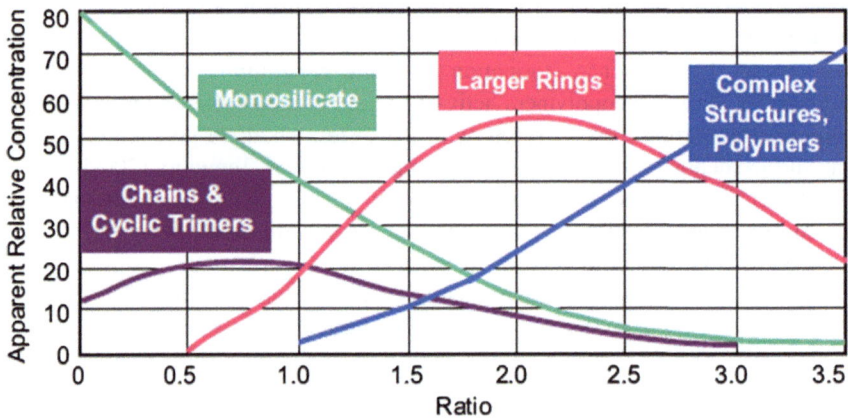

Fig. 2.1 Silicate anion structure equilibria (1 M solution) (PQ Corporation 2005) (The ratio is the modulus of the sodium silicate solution)

2.2 Alkaline Dissolution Media

can take up to 48 h to respeciate at room temperature. In very dilute solutions it can occur in a matter of minutes.

The above confirms the work of Applin (1987) who thought that at pH values in excess of 13 the predominant species was $SiO_2(OH)_2^{2-}$.

Kinrade and Pole (1992) working on zeolites thought that three main structural directing roles could be attributed to M^+. Firstly it stabilises oligmeric "mineral precursor" species in aqueous alumina-silicate systems. Secondly, hydrated M^+ have a templating effect whereby silicate and aluminate species replace water of hydration for a sufficient time to enable them to join together in a structural configuration. A third hypothesis was that the extent of condensation in alkali metal-silicate solutions increases slightly as the molecular weight of M^+ increases and M^+ forms ion-pairs with silicate anions. Higher silicate oligomers are favoured due to preferred ion-pairing with heavier cations.

They concluded that there are three means for alkali metal ions to influence aqueous silicate equilibria:

1. Electrostructive water structuring by small cations increases overall level of polymerisation. They thought that this was only relevant to lithium.
2. Ion pairing of silicate anions and M^+ increased polymerisation. Cations promote encounters between dissolved silicate anions by overcoming their electrostatic repulsion. Strongly paired cations will resist subsequent formation of a siloxane bond, thus the extent of polymerisation increases in the order $Li^+ < Na^+ < K^+ < Rb^+ < Cs^+$ (i.e. as the ion pair formation constants decrease).
3. Ion paired cations stabilise several specific oligomers by immobilising labile appendages and large ring structures.

Phair and van Deventer (2001) used ^{29}Si NMR to analyse the degree of polycondensation during geopolymerisation and confirmed that the basicity and hydration sphere of the alkali metal as well as it's concentration had a controlling effect on the degree of polycondensation and rate of dissolution from aluminosilicate minerals.

Sodium Aluminate (SA) is prepared by the action of sodium hydroxide solution on aluminium metal or by dissolving alumina in sodium hydroxide solution. Industrially aluminium hydroxide is dissolved in 20–25 % sodium hydroxide. Higher concentrations of sodium hydroxide give a semi solid product. In the production of solid SA aluminium hydroxide is added to boiling 50 % sodium hydroxide until a pulp forms. This pulp is cooled to give a 70 % solids SA product which is crushed and dehydrated to give a solid product (Kirk-Othmer 1991).

SA solutions tend to precipitate out alumina on standing unless stabilised with an excess of sodium hydroxide. Unstabilised SA solutions have a $Na_2O:Al_2O_3$ molar ratio of 1.51:1. Stabilised solutions have an excess of 11–13 % NaOH and a molar ratio in the range 1.75–1.95 (Orica Chemicals 2011). Total solids should also be considered with 38 wt% being usual (Southern Ionics 2006). However, the Coogee Chemicals (2012) stabilised product is a nominal 44.5 wt% solids with alumina content of 19 and 25.5 wt% NaOH.

SA may be obtained from the following waste product streams:

1. Aluminium industry as salt cake and spent potliner. Spent potliner is hazardous due to the presence of cyanides and fluorides and therefore requires pre-treatment to obtain SA. Salt cake is a by-product of aluminium dross formation in the recycling of aluminium metal. Processes to obtain SA from the salt cake recycling process are available (Rayzman et al. 1998).
2. Anodising bath spent solutions (Nugteren et al. 2011).
3. Bayer liquor from alumina refineries (Sipos 2009).

Aluminium does not appear to form products analogous to sodium and potassium water glass solutions even though it is present in the same tetrahedron configuration as silicon. This may be as a consequence of Lowenstein's rule. Catlow (1996) investigated the energies of formation and hydration of small clusters and rings containing silicon and aluminium. The results provided a rationalisation of Lowensteins rule in terms of the energetics of species containing Al–O–Al bridges. Whilst the formation of dimers is calculated to be energetically unfavourable in the unhydrated state, when the effects of solvation energies are added the reaction is now exothermic and possible. The formation of small rings involving Al–O–Al bridges is energetically unfavourable and hence further condensation of dimers is considered unlikely.

Moolenaar (1970) obtained vibrational spectra of sodium and potassium aluminates at a range of concentrations in sodium hydroxide solutions. Single crystal X-ray diffraction was used to determine the structure of potassium aluminate and demonstrated the presence of the dimer ion, $(OH)_3AlOAl(OH)_3^{-2}$, built up from two AlO_4^- tetrahedral which share an oxygen. At less than 1.5 M (aluminium) $Al(OH)_4^-$ was the main component. At 6 M an equilibrium between $Al(OH)_4^-$ and $Na_2Al_2O_4^{-2}$ (dimer) co-existed. Also in this work stable solutions containing appreciable amounts of aluminium could not be prepared at pH values less than 13.

Chen (1992) prepared SA solutions of equal concentrations by two different methods. Method 1 involved dissolving aluminium metal into highly concentrated NaOH solution. Method 2 involved dissolving aluminium metal in dilute NaOH solution, leaving it to stand for 15 days and solid NaOH added to bring the solution composition to that of Method 1.

Raman spectra of potassium and barium aluminate dimers show peaks at 540 and 550 cm^{-1}, respectively. The spectra from material prepared by method 1 showed a higher peak in this region than did method 2 material. The authors believe that this is indicative of higher sodium aluminate dimer content and this is supported by higher solution viscosity and lower electroconductivity in material from method 1 compared to material from method 2.

Nianyi and Honglin (1994) studied the aluminate ion using quantum chemical methods. The Al–O bonds (of the Al–O–Al bridge) in aluminate ions have significant covalent character so reactions involving the formation or cleavage of the Al–O bond may be a slow process. In UV spectra a broad absorption band near 2,700 Å was assigned to the aluminate dimer ion. Excitation energy calculations gave a value of 2,666 Å close to the experimental value. Excitation energy calculations for Al(OH)$_4^-$ and Al(OH)$_6^{3-}$ gave wavelengths of 2,300 and 2,645 Å, respectively.

Concentrated sodium aluminate solutions with high Na$_2$O:Al$_2$O$_3$ ratio contain small amounts of Al(VI) as Al(OH)$_6^{3-}$. Ma et al. (2007) attributed a UV peak at 3,700 Å to this Al(VI) co-ordinated species. They claim that this peak only appears when solutions are made by dissolving aluminium trihydroxide. This peak disappeared when the solution was heated to 135 °C.

Bands observed at 725 cm^{-1} (infrared) and 621 cm^{-1} (Raman) are attributed to asymmetric and symmetric Al–O stretching vibrations of the aluminate ion, Al(OH)$_4^-$ in solution. Other low intensity bands at 890 and 630 cm^{-1} (infrared) and 540 cm^{-1} (Raman) indicate the formation of a second aluminate ion species, the dimer.

Sipos (2009) reviewed the structure of aluminium in strongly alkaline aluminate solutions. Apart from Al(OH)$_4^-$ only one further Raman distinguishable aluminate species appears to exist in significant concentrations, the dimer.

One possible explanation of unusually high viscosity values of SA solutions is that aluminate ions in aqueous solutions form an extended network of ions (both same and opposite charge) which is held together by bridging counter ions and possibly hydrogen bonds. Such extended networks could explain the variations observed in transport properties of these solutions. This network cannot be described as a polymer, but as a diffuse network of hydrated closely packed ions.

Solutions of SA appear to contain monomer, dimer and some aluminium in VI co-ordinate form. Therefore care must be taken in identifying the composition of SA solutions prior to use in geopolymer synthesis.

Brew and MacKenzie (2007) produced a geopolymer by reacting sodium aluminate solution with a slurry of silica fume in sodium hydroxide. By keeping the H$_2$O:Na$_2$O ratio <10 an X-ray amorphous material was obtained. Higher ratios gave rise to crystalline zeolites.

^{27}Al MAS NMR of the sodium aluminate solution gave a sharp resonance at 78.3 ppm indicative of crystalline material containing aluminium in tetrahedral co-ordination. There is evidence of a very small aluminium resonance at 13 ppm indicative of octahedral aluminium. This latter peak remained in the geopolymer made from the sodium aluminate. Curing for 48 h at 40 °C gave a compressive strength of 26 MPa.

Nugteren et al. (2011) investigated waste etching solutions as an alkaline source to activate fly ash and slag to manufacture geopolymers. Potassium silicate/potassium hydroxide/sodium aluminate blends were evaluated and compared to waste etch bath solution. Potassium silicate/potassium hydroxide activation gave a compressive strength of 94 MPa. Gradually replacing potassium hydroxide with sodium aluminate gave a dramatic drop in compressive strength to 3 MPa. This was attributed to preferential reaction of the potassium silicate and sodium aluminate with little waste activation occurring. Additional water was added to improve workability which would also have inhibited strength development. Using sodium aluminate as the sole activator saw compressive strength increase to 31 MPa, whilst the addition of the etch bath activator (polyalcohol/thiosulphate) at 0.4 wt% saw a further increase to 38 MPa.

The use of the waste solution in combination with potassium hydroxide gave a range of compressive strengths between 43 and 59 MPa. Replacing potassium hydroxide with sodium hydroxide gave a drop in compressive strength from 59 to 49 MPa. Increasing the fly ash to slag ratio (reduced calcium) saw the compressive strength drop from 59 to 43 MPa.

The presence of calcium containing slags leads to the formation of calcium aluminate hydrates in which aluminium is present in VI co-ordinate state.

The authors commented that when excess silica (from potassium silicate) is available in solution, liberation of alumina may quickly result in the formation of sialate chains to form geopolymer precursors. These pastes will thicken and the setting point will be quickly reached. However, when aluminate is present instead of silica, it will take time before sufficient silica is liberated from the fly ash to give a suitable silicon to aluminium ratio to form geopolymer gels. It will take longer to thicken and set.

Phair and van Deventer (2002) investigated alkali activation of various materials with several combinations of alkali activating solutions. In the case where SA ^{27}Al NMR and FTIR were performed on the sodium aluminate solution, they found that the sodium aluminate solution they used to be predominantly IV co-ordinate with a small amount of VI co-ordinate material. (Note: the SA specification was $Al_2O_3{:}Na_2O = 0.89$ and 26.5 wt% Al_2O_3. This is an excess of around 12 wt% alkali). FTIR of the geopolymer samples revealed a peak at 730 cm^{-1} indicating VI co-ordinate aluminium for 5.1 M aluminium. Peaks for IV co-ordinate aluminium were less prominent. Increasing the NaOH content has a minimal effect on reducing the VI co-ordinate aluminium or promoting the presence of IV co-ordinate material.

Bayer liquor is a by-product from alumina refineries. A typical composition is shown in Table 2.3 (Marciano et al. 2006).

Bayer liquor (Jamieson et al. 2012) has been used to replace sodium silicate as an activator for fly ash based geopolymers. Compressive strengths in excess of 40 MPa were obtained for geopolymer pastes cured at ambient temperature.

Verdolotti (2008) compared NaOH and sodium aluminate slurry in NaOH to consolidate pozzolanic soils. They claim that the IV co-ordinate aluminium atoms present

Table 2.3 Typical Bayer Liquor Composition based on Marciano et al. (2006)

Component	Concentration
Al_2O_3	96 g/l
Na_2O caustic	160 g/l
Na_2O carbonate	28 g/l
Organic carbon	10 g/l
Cl$^-$	3 g/l
SiO_2	1 g/l
SO_3^-	0.7 g/l
pH	~14
TDS	30 wt%
Viscosity at 50 °C	3,300 Pa.s

in the pozzolana and sodium aluminate change their co-ordination state splitting between IV and VI co-ordinate, modifying the traditional polysialate chemistry.

The pozzolana had silicon to aluminium compositional ratio of 3:1; this ratio in the geopolymer was 2.71:1 for 10 M NaOH and 2.28 for the sodium aluminate slurry in NaOH. This would indicate that the aluminium from the sodium aluminate is affecting the final composition.

The sample treated with 10 M NaOH showed the highest concentration of Q^4 (1Al). The presence of sodium aluminate promotes the co-ordination of four aluminium atoms to SiO_4 tetrahedra. SEM micrographs showed amorphous morphology from 10 M NaOH activated samples. The sample activated by sodium aluminate showed a less homogeneous structure plus needle like crystals spreading over the sample surface.

XRD patterns (Cu kα radiation) showed new crystalline peaks at 42° (2θ) in both geopolymer samples and a peak at 27.5° (2θ) in the sodium aluminate activated sample. Neither NaOH nor sodium aluminate have reflections at these positions. The peak at 42° (2θ) can possibly be attributed to zeolite formation. The intensity of the peak at 27.5° (2θ) reduced during the curing period. The authors, supported by FTIR and NMR data, concluded that sodium aluminate activation was capable of generating VI co-ordinate aluminium at least during short term curing.

The presence of Al (VI) co-ordinated aluminium species, for short curing times, might affect the polysialate formation in terms of aluminium type and quantity available.

"This result represents further evidence that the polysialate chemistry based on IV co-ordinate aluminium is not capable of describing all the phases of geopolymers".

The best compressive strengths, 20 MPa (10 M NaOH) and 45 MPa (10 M NaOH/sodium aluminate), were obtained after 300 days curing time.

Hajimohammadi et al. (2008) described the development of a "just add water" geopolymer cement based on sodium aluminate (solid) and geothermal silica. He observed that high early aluminium concentration inhibits geopolymerisation whilst high early silicon concentration enhances the reaction. To match these requirements the silica source must have a fast dissolution rate so that solutions will rapidly become rich in silicon entities similar to sodium silicate solutions. High early concentrations of aluminate are adsorbed onto the silica surface inhibiting dissolution and high early silicon forms silicate oligomers which can react with aluminate promoting geopolymerisation.

2.3 Admixtures and Fillers for Geopolymer Systems

2.3.1 Admixtures

Additives to geopolymer systems can be used to enhance specific properties of the system. The term admixture (ASTM Committee C09 2012b) is used to cover the wide range of materials added to OPC.

Improvements in workability and setting, reduction in final density, improvements in physical properties, particularly toughness and thermal performance can all be enhanced by the judicious choice of additive(s). The economics of geopolymer systems can be influenced by the addition of a wide range of fillers (aggregates) of varying compositions and size. Materials such as fly ash, fumed silica, metakaolin and slag are considered as admixtures in OPC (ASTM Committee C09 2012a), but in geopolymers they are considered to be part of the binder system.

Test methods for workability have been reviewed (Koehler 2009; Koehler and Fowler 2003). Kantro (1980) developed the mini cone slump test which is suitable for evaluation of the addition of admixtures to cementitious pastes.

In OPC processing superplasticisers are used to improve workability (and reduce water demand) and there are several classes available commercially (Table 2.4) (Rixom and Mailvaganam 1999).

Superplasticisers are polymeric molecules with several anionic segments. When anionic superplasticisers are added to cement the negative charged segments adsorb on to the surface of the cement particles increasing the overall negative charge. This causes mutual repulsion of the cement particles, breaking up the flocculation and releasing trapped water. This released water reduces the plastic viscosity, improving workability and hence the water-cement ratio can be reduced. In addition to electrostatic repulsion, superplasticisers containing non-polar polymer segments also function by steric hindrance. Segments of the polymer protrude out from the polymer surface into the aqueous phase preventing cement particles from agglomerating (BASF 2008).

Table 2.4 Superplasticiser structures adapted from Rixom and Mailvaganam (1999)

Class	Origin	Structure (typical repeat unit)
Lignosulphonates	Derived from neutralization, precipitation, and fermentation processes of the waste liquor obtained during production of paper-making pulp from wood	
Sulphonated melamine formaldehyde (SMF)	Manufactured by normal resinification of melamine-formaldehyde	M=Na
Sulphonated naphthalene formaldehyde (SNF)	Produced from naphthalene by oleum or SO_3 sulphonation; subsequent reaction with formaldehyde leads to polymerization and the sulphonic acid is neutralized with sodium hydroxide or lime	R=H, CH_3, C_2H_5; M=Na; SNF
Polycarboxylic ether (PCE)	Free radical mechanism using peroxide initiators is used for polymerization process in these systems	EO: Ethylene oxide

2.3 Admixtures and Fillers for Geopolymer Systems

There are many references to the use of superplasticisers in OPC (Liu et al. 2008; Xiao et al. 2011; Bassioni 2010; Wild et al. 1996a, b; Yilmaz et al. 1993; Jiang et al. 1999; Gołaszewski and Szwabowski 2004; Fernàndez-Altable and Casanova 2006a, b; Yllmaz et al. 1993; Agullo et al. 1999).

References to the use of superplasticisers in geopolymer are more limited and are generally based on slag based systems. Bakharev (2000) investigated a range of admixtures in alkali activated slag concrete and found that the air entraining agent improved workability. She suggested that admixtures with nonpolar molecules are better suited to work in a media of strongly charged particles of alkali activated slag and activator. Palacios (2005) found that only a naphthalene based product retained paste fluidity in a sodium hydroxide activated paste. This was attributed to the good sodium hydroxide resistance of this admixture. The use of calcium hydroxide solution in the chemical resistance test did not affect any of the tested admixtures and confirmed their suitability for use in OPC.

Yilmaz et al. (1993) treated SMF and SNF superplasticisers in 1 M KOH solution. The SMF precipitated out of solution, whilst the SNF was substantially unaffected.

Work by Hardjito (2005) on fly ash based geopolymer concrete used a naphthalene sulphonate type super plasticiser at up to 4 wt% and achieved increased slump (Fig. 2.2). Above 2 wt% super plasticiser addition compressive strength began to decline.

Ghosh and Ghosh (2012) evaluated the workability of fly ash based geopolymer mortar using a mini slump cone. The effect of alkali content, silicon to aluminium ratio, water content and plasticiser dosage where evaluated. He considered that 150 mm flow diameter was the minimum value suitable for mortar that could be readily placed in a mould. Table 2.5 shows their suggested workability criteria for geopolymer mortars.

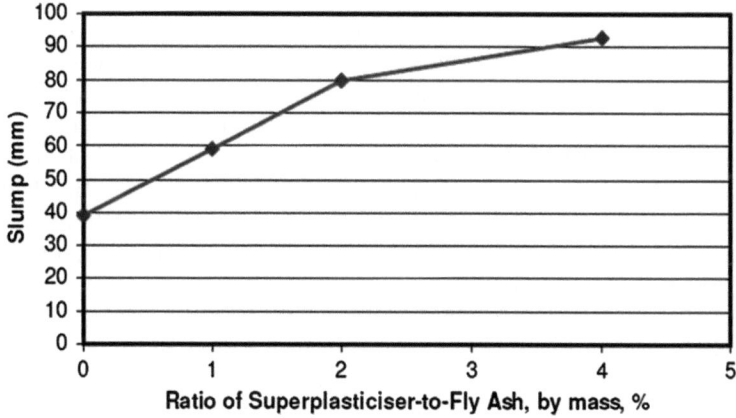

Fig. 2.2 Geopolymer concrete slump versus superplasticiser content (Hardjito and Rangan 2005)

Table 2.5 Workability criteria for geopolymer mortar (Ghosh and Ghosh 2012)

Flow diameter, mm	Workability
Above 250	Very high
180–250	High
150–180	Moderate
120–150	Stiff
Below 120	Very stiff

Increasing alkali content and silica content reduced workability. Dosages of naphthalene super plasticiser above 1 wt% gave marked improvements in workability.

Memon (2012) reported on the production of self-compacting geopolymer concrete. This type of concrete must be able to flow into position without the use of external vibration. They found that the addition of superplasticiser up to 7 wt% not only improved flowability but increased compressive strength.

Moura (2011) investigated the formulation of metakaolin geopolymer repair mortars. He found that the use of 3 wt% superplasticiser with 10 wt% calcium hydroxide and using 10 M sodium hydroxide gave 80 % increase in flow. Compressive strengths over 40 MPa were obtained.

Setting times can be adjusted by the use of accelerators and retarders. Calcium oxide and hydroxide have been reported to accelerate geopolymer setting (Jamieson et al. 2012; Temuujin et al. 2009b).

In US patent 7794537 (Barlet-Gaudedard et al. 2010) the use of lithium chloride, up to 7 %, acts as an accelerator, whilst 0.65 wt% sodium pentaborate extended setting time from 1.5 to 3 h at 110 °C. Nicholson (2005) claimed the use of inorganic boron compounds for retarding the setting time of high calcium fly ash based geopolymers.

The introduction of gas cells into geopolymers gives rise to materials with good sound and thermal insulation properties. Air can be entrained in the system during the mixing process and the air bubbles are stabilised by the addition of air entraining admixtures, which reduce the surface tension at the air-water interface. Alternatively gas bubbles can be generated in situ by chemical reaction. Blowing agents can be hydrogen peroxide, sodium perborates and the reaction of aluminium metal with the activating alkali solution (Liefke 1999; Reid 2011; Vaou et al. 2010). Bell and Kriven (2009) compared hydrogen peroxide and spherical aluminium powder as blowing agents in metakaolin based geopolymers. The hydrogen peroxide gave a non-percolating foam structure which cracked on firing. The aluminium powder gave irregular shaped pores with a wide pore size distribution which was crack free after firing.

Cementitious materials, typically concrete, are the most widely used materials for infrastructure construction. They are typically characterised by low tensile strength and low strain capacity and are sensitive to micro cracking. Fibres and steel and FRP rebar may be incorporated into cementitious matrices to overcome these deficiencies giving materials with increased tensile strength, ductility, toughness and increased durability (Chanh 2004; Hameed et al. 2009).

2.3 Admixtures and Fillers for Geopolymer Systems

Steel bars are the materials of choice for structural applications, but the addition of "short fibres" now plays an important role in the processing and the development of early stage properties such as plastic and drying shrinkage. The addition of non-metallic structural fibres in geopolymers is opening up new applications. Table 2.6 compares the three main classifications of discontinuous fibres. Additionally microfibers can reduce spalling in fire situations. The cement and concrete Association of New Zealand has published a bulletin on fibre reinforced concrete (CCANZ 2009).

Natural fibres are also finding utility in concrete (Buckeye Building Fibres 2009; Hercules Fibres 2011b). Both the natural cellulose fibres listed here are shorter than typical polypropylene fibres of the same diameter. This will lead to more fibre per unit volume of fibre and hence the possibility of better crack control. These shorter fibres should have less influence on workability.

Table 2.6 Comparison of steel and synthetic fibres (Wimpenny et al. 2009)

	Steel fibres	Synthetic micro-fibres	Synthetic macro-fibres
Property	Characteristic property of fibres		
Shape/texture	Cold drawn hooked ends	Straight smooth	Continuously embossed
Collation	Glued bundles	Fibrillated	Uncollated
Typical length (mm)	60	12	48
Typical diameter (mm)	0.75	0.02–0.03	0.5–1
Tensile strength (MPa)	1050	30	550
Elastic modulus (GPa)	>200	2	10
Dosage (kg m^{-3})	25–35	1–2	6–10
Service temperature (°C)	300	60	60
Melting point (°C)	>800	150	150
Base material	Carbon steel	Polypropylene	Polyolefin (polypropylene, polyethylene)
Comparison with conventional concrete (unreinforced except for *)			
Workability	Reduced	Slightly reduced	Slightly reduced
Plastic shrinkage cracking	Unaffected	Reduced	Slightly reduced
Early-age thermal cracking	Reduced	Unaffected	Reduced
Long-term shrinkage cracking	Reduced	Unaffected	No data
Stray current corrosion	Reduced	Unaffected*	Eliminated

(continued)

Table 2.6 (continued)

	Steel fibres	Synthetic micro-fibres	Synthetic macro-fibres
Durability in chloride exposure*	Increased	Unaffected*	Greatly increased
Fire spalling resistance	Slightly increased	Greatly increased	Increased
Compressive strength	Unaffected	Unaffected	Unaffected
Residual flexural strength	Increased	Unaffected	Increased
Impact strength	Greatly increased	Unaffected	Increased
Flexural toughness	Increased	Unaffected	Increased
Abrasion resistance	Increased	Slightly increased	Slightly increased
Freeze-thaw resistance	Slightly increased	Increased	Increased
Flexural energy absorption	Greatly increased	Unaffected	Greatly increased
Concrete permeability	Slightly increased	Slightly increased	Slightly increased
Pump wear	Increased	Reduced	Reduced
Safety*	Hazard from handling and protruding fibres	Increased	Increased
Finishing	Extra care during floating	Exposed fibres soon abrade	Fibres may float and protrude in poorly designed mixes

For high temperature applications a range of fibres with increased thermal stability is available. These include, but are not limited to, alkali resistant glass fibre, basalt fibre, carbon fibre, ceramic fibres (alumina and silicon carbide) and wollastonite (Thang et al. 2010; Fibres Unlimited 2007; Morrison 2008; Silva and Thaumaturgo 2003; Chen and Chung 1993).

Fibre blends have been introduced to overcome short comings in the use of single type/size fibres. A blend of long and short fibres, where the long fibre imparts structural improvements and the short fibre imparts shrinkage improvements is typical. These blends can be all steel fibres, all organic fibres or mixtures of the two (CCANZ 2009).

Figure 2.3 and Table 2.6 show the types and properties of the two most commonly used fibre types i.e. steel and polypropylene.

2.3.2 Fillers

Fine aggregates (<53 μm) can be used as fillers to improve specific properties. Wollastonite, chamotte, and α-alumina are all stable beyond 1,000 °C and help to control shrinkage in high temperature applications (Kamseu et al. 2010; Buchwald et al. 2009).

2.3 Admixtures and Fillers for Geopolymer Systems

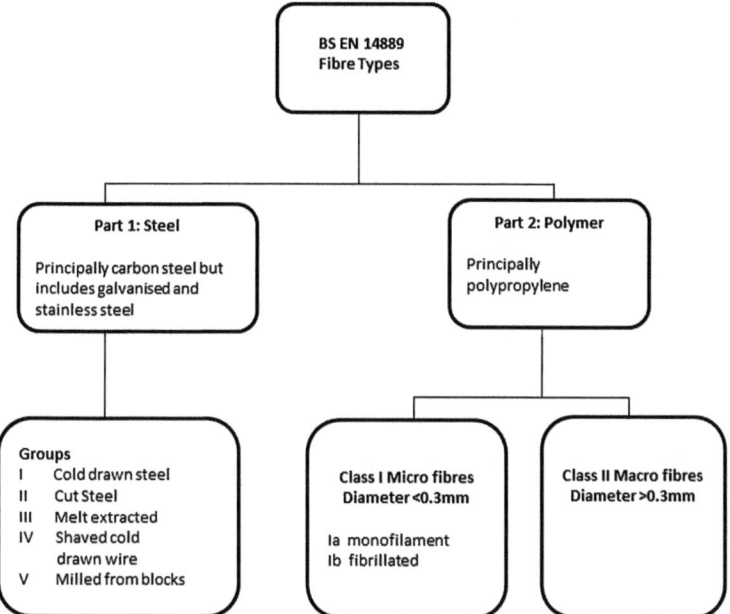

Fig. 2.3 Fibre types according to BS EN 14889 (Wimpenny et al. 2009)

Davidovits (2008a) compared cordierite (magnesium aluminosilicate) and mica in K-polysialate geopolymer and in a (Na, K) polysialate and measured thermal behaviour after an initial post cure of two hours at 650 °C. In the second heating cycle the cordierite filled (Na, K) polysialate showed an expansion of 0.1 % up to 600 °C before undergoing strong shrinkage above this. The mica filled system showed 0.5 % expansion up to 600 °C before shrinking. Davidovits attributed this shrinkage to the collapse of the sodalite cage. A Na polysialate filled with cordierite and post cured for 3 h at 500 °C gave an average coefficient of thermal expansion (COTE) between 20 and 700 °C of $1.9 \times 10^{-6} \, K^{-1}$.

Barbosa and MacKenzie (2003b) added various inorganic fillers to a metakaolin based geopolymer activated with sodium silicate and measured physical properties and thermal shrinkage of the composites. The fillers included crushed building brick, powdered granite, kaolinite, iron sand (black titanomagnetite with 7 % TiO_2), α-alumina and β-Sialon. All the filled geopolymer composites showed reductions in strength which was attributed to the lack of reaction between the fillers and the geopolymer binder.

Lin et al. (2009b) investigated the use of alumina as a filler in a metakaolin based geopolymer and found marked reductions in shrinkage after elevated temperature exposure.

Siva and Thaumaturgo (2003) used wollastonite to improve fracture toughness of metakaolin based geopolymers. The addition of 2 vol.% of wollastonite gave the optimum toughening effect.

Temuujin et al. (2012) added ground vermiculite to a fly ash based geopolymer to improve the thermal properties of a sprayed coating. Vermiculite expands when heated above 300 °C showing low thermal conductivity. The ground sizes used were −63 and −250 μm. The presence of vermiculite modified the water release mechanism on heating. The early water loss shown by unfilled geopolymers up to 400 °C is replaced by a more continuous loss up to 1,000 °C. This latter effect has a more beneficial influence on fire resistance.

Mullite is usually present in coal fly ash and remains reasonable stable through the geopolymerisation process and any elevated temperature exposures up to 1,000 °C (Williams and van Riessen 2010; Rickard et al. 2012). This probably contributes to the thermal stability of fly ash based geopolymers. Mullite is available commercially (Washington Mills 2008) and is used commercially in the manufacture of refractory products where resistance to spalling and low thermal conductivity are important. Mullite also possesses a low (5.62×10^{-6} K^{-1}) COTE between 25 and 1,500 °C without any phase changes occurring in this range.

Medri et al. (2011) used a potassium activated metakaolin as a binder for recrystallised silicon carbide to produce refractory coatings for silicon nitride-titanium nitride substrates. After curing at 80 °C and subsequent water removal a coating with 90 wt% silicon carbide was obtained. Long term oxidative testing of the silicon carbide grains and the filled geopolymer coating was carried out for 100 h at 1,200 °C. The weight gain of the silicon carbide coating in this test was 7 % and is around 50 % lower than for the grains. Shrinkage of these coatings after 1 h at 1,300 °C in argon was 0.1 %. Adhesion of the coating is good as shown by the absence of exfoliation detachment when cutting or polishing the cross sections prior to and after high temperature exposure.

Wang et al. (2011) added hollow cenospheres sourced from coal fly ash to geopolymers to reduce density and reduce thermal conductivity in the composites. The geopolymer was a potassium activated metakaolin and the cenospheres were claimed to have good alkali resistance. The addition of 40 vol.% of cenospheres reduced compressive strength from 105 to 40 MPa, the thermal conductivity from 0.361 to 0.173 W m^{-1} K^{-1} and density from 1.37 to 0.82 g cm^{-3}.

Chen et al. (2011a) developed a sodium activated metakaolin geopolymer suitable for coating concrete where good thermal insulation properties were required. The addition of mica (15 %) and wollastonite (10 %) together with 5 % of a silane gave a product with good compressive strength (27.4 MPa) and adhesion (5.14 MPa). To reduce the thermal conductivity hollow glass spheres (20–50 μm diameter) were added to the filled geopolymer. A 4:1 by volume of glass spheres to geopolymer mix gave the optimum results with respect to compressive strength (18.9 MPa), adhesion (4.78 MPa) and thermal conductivity (0.21 W m^{-1} K^{-1}).

Concrete is a cement-water paste in which aggregate is embedded. Aggregate is granular material such as sand, gravel, crushed stone, and other speciality materials that occupy 60–75 % of the concrete volume. Aggregate properties affect the plastic concrete (workability and drying shrinkage) and hardened properties such as durability, strength, thermal properties and density.

2.3 Admixtures and Fillers for Geopolymer Systems

Aggregates are classified as natural, synthetic and recycled. Grading is by particle size and size distribution, particle shape and water content. Crushed aggregate obtained by quarrying and crushing of rock can contain substantial amounts of flat or elongated particles, whereas natural gravel has a rounded shape and smooth surface texture (Cement Concrete and Aggregates Australia 2008). The ACI committee E-701 produced a bulletin on "Aggregates for Concrete" (ACI E-107 2007).

Chapter 3
Chemistry of Geopolymers

Abstract This chapter focuses on how the aluminosilicate sources react with the alkaline activating solutions. The application of model systems based on metakaolin is explored in detail and a simplified geopolymerisation model is shown. The critical influence of water content on the outcome of geopolymerisation is clearly demonstrated. The wide range of analytical techniques employed to characterise the precursors and geopolymer microstructure is addressed. In the case of fly ash based geopolymers the interpretation of kinetic and mechanistic results is more difficult due to the presence of crystalline materials. Variation in fly ashes sourced from different plants also contributes to these difficulties. The alkali reactive glassy aluminosilicates may not be readily accessible for alkali dissolution reactions, being shielded by non-reactive phases. An alkali activation of fly ash model is described. The presence of soluble calcium entities changes the kinetics of fly ash geopolymerisation. It is shown that increasing CaO content generally increases compressive strength, reduces setting times and facilitates ambient curing.

Keywords Model systems · Alkali activation · Water content · Amorphous content of fly ash · Calcium compounds

Essentially geopolymer formation is by alkali dissolution of an amorphous aluminosilicate material, followed by hydrolysis of dissolved Al^{3+} and Si^{4+} entities and condensation of the resulting silicates and aluminates to form a geopolymer (Weng and Sagoe-Cretsil 2007).

The complexities of the geopolymerisation reaction demand that a model compound be used to elucidate the mechanisms involved. Metakaolin is the material of choice by many research groups (Weng and Sagoe-Cretsil 2007; Rahier et al. 1996; Davidovits 1991; Barbosa et al. 2000b; Alonso and Palomo 2001; Rowles and O'Connor 2003).

3.1 Metakaolin Based Geopolymers

Metakaolin reacts with alkaline sodium silicate solution at temperatures below 100 °C to produce an amorphous, glassy geopolymer (Rahier et al. 1996). Thermal analysis (DSC and TGA), XRD, NMR and mechanical testing were used to characterise the system. The exothermic reaction of metakaolin and sodium silicate solution was followed by DSC. This showed that realistic isothermal cures could be obtained at 60 °C. The maximum heat flow is obtained at the beginning of the reaction and after 4 h approximately 83 % of the maximum reaction enthalpy had been realised. The maximum reaction enthalpy was determined to occur when the metakaolin to sodium silicate molar ratio was 1.0.

XRD scans show that metakaolin is predominantly amorphous apart from traces of titanium dioxide impurity from the source kaolin. The resulting geopolymer is also x-ray amorphous.

The ^{27}Al MAS NMR spectrum for the geopolymer formed shows a single peak at 58 ppm with a FWHM of 16 ppm. This is indicative of tetrahedral aluminium surrounded by 4 silicon tetrahedra. Since only a signal for aluminium tetrahedra is observed in the geopolymer, the other aluminium sites present in metakaolin (Al(V) at 35 ppm and Al(VI) at 0 ppm) are transformed into Al(IV). The same transformation is observed when sodium and potassium hydroxides are used to react with metakaolin. Since the aluminium is four co-ordinated, a sodium ion is required for each aluminium ion for charge balancing and hence the 1:1 sodium to aluminium ratio seen in the DSC results (Note: If aluminium is already present in the raw material precursor as Al(IV), then it must already have a charge balancing cation associated with it and will not require M^+ from the activating solution. Sodium to aluminium ratios lower than unity can be used in these cases to prevent excess alkalinity which causes carbonation and efflorescence).

Compression testing also showed a maximum value of 60 MPa at a sodium to aluminium ratio = 1.

Rahier concluded that the silicate and aluminate monomers are combined in a random way but with the restriction that each aluminium centre must be linked to four SiO_4 groups so that no Al-O-Al are present. The silicon to aluminium ratio in the geopolymer cannot be influenced by the variability in the reaction mixture but must conform to a unique stoichiometry (Note: excess SiO_2 can form water soluble silicate polymers, which can be dissolved out, as shown by before and after leaching NMR spectra, but these polymers can also graft on to the 1:1 stoichiometric geopolymer). He also concluded that less than 5 wt% of the total water in the reaction mixture was chemically bound as Si-OH.

Barbosa et al. (2000b) produced geopolymers from metakaolin (calcined for 6 h at 700 °C) and sodium silicate (modulus = 2.0) solutions. The silicon to aluminium ratio and sodium to silicon ratios were varied as was the water content of the reaction mixtures.

The viscosity of one sample ($SiO_2:Al_2O_3 = 3.3$, $Na_2O:SiO_2 = 0.25$, $H_2O:Na_2O = 10$), which gave the best compressive strength of 48 MPa after 1 h cure at 65 °C was

3.1 Metakaolin Based Geopolymers

monitored over time. When the results were plotted on a log-log scale a marked change in slope after 40 min was observed. This is analogous to increases in melt viscosity as molecular weight increases in organic polymers. The authors proposed that at shorter times before the change point the growing oligomers behave as a dilute solution. As the concentration of oligomers grows they become closer together and are able to react increasing the molecular weight which shows as gel formation beyond the point of slope change.

Samples with $H_2O:Na_2O$ ratios of 25 did not cure sufficiently at 65 °C and could not be tested. Samples with a $H_2O:Na_2O$ ratio of 10 gave testable samples, which decreased in compressive strength as the $SiO_2:Al_2O_3$ and the $Na_2O:SiO_2$ ratios increased.

MAS NMR spectra were collected on well cured and under cured samples in the liquid phase. ^{29}Si spectra for the well cured sample showed an immediate broadening of the peak and a shift to less shielded values when metakaolin was added to the alkaline activating solutions suggesting an immediate reaction between silicate solution and aluminate in metakaolin. As this mix was monitored the silicate peaks diminished in intensity over the first 7 h.

In the case of under cured samples peaks broadened on initial addition of metakaolin, suggesting an initial reaction, but the silicate peak intensities remain virtually unchanged up to 15 h after mixing, indicating that consumption of the silicate by polymerisation is not proceeding.

FTIR results showed the presence of sodium carbonate in the under cured samples, but not in the fully cured. The sodium carbonate is formed by reaction of sodium silicate and/or sodium hydroxide with atmospheric carbon dioxide. Attempts to detect the presence of a ^{27}Al NMR signal were unsuccessful indicating insignificant dissolution of aluminium from the metakaolin.

MAS NMR of ^{27}Al in the geopolymer showed a strong, narrow peak in the well cured geopolymer, but the under cured samples showed intensities of the same peak ranging from 3.4 to 11 %. This indicates that the aluminate tetrahedra are more ordered in the well cured geopolymer.

The optimum conditions for geopolymer formation are when $Na_2O:SiO_2 = 0.25$, and $H_2O:Na_2O = 10$. This sodium level satisfies the charge balancing requirements without excess material for carbonation reactions which can disturb polymerisation reactions. Sufficient water is required for wetting and mixing and to provide a mechanism for ionic transport. The effect of extra water may be to dilute the reaction (reduce pH) and transport soluble species away from reaction zones.

Rahier (1996) had shown previously that the aluminium to silicon ratio in the geopolymer is the same as aluminium to silicon ratio in the precursors if M^+:aluminium $= 1$. He then went on to investigate the reactions between sodium and potassium silicate solutions and metakaolin (Rahier et al. 1997). The influence of the molar ratios $H_2O:R_2O$ (w) (between 6.6 and 21) and $SiO_2:R_2O$ (s) (between 0 and 2.3) of the silicate solution were investigated.

Differential Scanning Calorimetry (DSC) could not be used for $w = 6.6$ (Na) and $w = 8.1$ (K) because of poor mixing. Poor wetting of the metakaolin particles at low w values could have led to incomplete reaction with corresponding lower enthalpy values. The effect is more pronounced with sodium silicates due to a higher solution viscosity compared to potassium silicate solutions.

Rahier (1997) Showed that when the $SiO_2:Na_2O$ ratio was greater than 0.8 amorphous geopolymer was formed. When the ratio was less than 0.8 the product was partially crystalline. This was attributed to depolymerisation of the larger silicate entities by the alkali cation.

The upper limit of the molar ratio, s, is also determined by the reaction rate which becomes slower with increasing s. For values above 2.3 for sodium silicate and 1.9 for potassium silicate the reaction does not go to completion within the time scale of days at ambient temperature. The reaction of metakaolin with potassium silicate is slower than for sodium silicate solutions with the same s and w values.

Zuhua (2009a) activated metakaolin with equal volumes of increasing molarity sodium hydroxide solution. He used calorimetry to measure the heat evolved as the molarity and time increased. There is an initial peak, the size of which increases with molarity. The author attributed this region to dissolution-hydrolysis. The heat evolution then decreases before increasing again. The rate of heat evolution and the maximum value of the second peak are higher for increasing molarity up to 9 M. Using 12 M NaOH led to a lower rate of heat evolution and reduced second peak height. The high hydroxyl ion concentration is postulated to be having an adverse effect on the polycondensation reaction, possibly by hydrolysis of the formed condensed products. This region is attributed to hydrolysis-polycondensation. The 3 M NaOH solution showed very little heat evolution in either region and this was attributed to the high water content of this solution.

The next series of calorimetry work used 12 M NaOH at liquid to solid ratios of 1, 1.2 and 1.25. The higher liquid to solid ratio showed the highest reaction in the first region, but in the second region the mechanism changes from hydrolysis (using water) to polycondensation (releasing water) and the lower liquid to solid ratio showed the highest heat evolution.

Zuhua believes that water is a reactant in the initial region (hydrolysis) and the hydroxyl group is the catalyst (Note: Purdon stated that alkali was a catalyst in activation of slag). He gives a series of reaction equations to support this position. Similarly water is a product of the polycondensation reaction so lower water contents at this stage will help the reaction move from left to right (towards products) to establish equilibrium. This product water is now available to take part in further dissolution reactions which are taking place simultaneously with the polycondensation.

Weng (2007) reported that Henry (1992) demonstrated that the chemical reactivity of metal ions during the processes of hydrolysis and polycondensation can be predicted and explained by the partial charge distribution in molecular species. The values calculated are useful only for direct comparisons.

The initial step in the geopolymerisation mechanism is the dissolution of metakaolin which provides aluminate ions and, at least, part of the silicate ions. Based on the partial charge method (PCM) the calculated results indicate that $[Al(OH)_4]^-$ is the major Al species under alkaline conditions, while the major

3.1 Metakaolin Based Geopolymers

silicate structures are $[SiO(OH)_3]^-$ and $[SiO_2(OH)_2]^{2-}$. The concentration ratio of $[SiO_2(OH)_2]^{2-}$ to $[SiO(OH)_3]^-$ increases with alkalinity. This can be depicted as follows:

$$Al_2O_3 + 3H_2O + 2OH^- = 2[Al(OH)_4]^- \qquad (3.1)$$

$$SiO_2 + H_2O + OH^- = [SiO(OH)_3]^- \qquad (3.2)$$

$$SiO_2 + 2OH^- = [SiO_2(OH)_2]^{2-} \qquad (3.3)$$

This suggests that H_2O and OH^- are consumed during dissolution.

The setting and hardening of geopolymers occurs as a result of condensation between aluminate and silicate species and this reaction is more rapid than the reaction between silicate species.

Equation 3.4 shows a molecule of water being released during the condensation reaction.

$$[Al(OH)_4]^- + [SiO(OH)_3]^- \rightarrow \begin{array}{c} H \\ HO \diagdown \diagup O-SiO(OH)_2^- \\ Al \\ HO \diagup \mid \diagdown OH \\ OH \end{array} \rightarrow \left[\begin{array}{c} O \; SiO(OH)_2 \\ \mid \\ HO-Al-OH \\ \mid \\ OH \end{array} \right]^{2-} + H_2O \qquad (3.4)$$

The aluminate can react with either of the silicate species shown above (Eqs. 3.2 and 3.3) and the authors show the range of stabilities of the various resultant products. To summarise the condensation between aluminate and silicate species is dependent on the speciation of the silicate entities. The condensation between $[Al(OH)_4]^-$ and $[SiO_2(OH)_2]^{2-}$ tends to form small oligomers such as dimers and trimers, whilst the condensation with $[SiO(OH)_3]^-$ results in larger oligomers and polymers. Therefore the formation of geopolymer depends on the concentration of $[SiO(OH)_3]^-$ which favours lower alkalinity values.

They proceeded to evaluate the influence of increasing sodium hydroxide molarity on silicon and aluminium entity dissolution from metakaolin and also carried out calorimetry. The calorimetry showed an immediate large exotherm effect when metakaolin was mixed with sodium hydroxide solutions in the range 1–15 M. A second exotherm evolved in the 5 M and 10 M solutions with that in the 10 M being faster and more pronounced. Dissolution of metakaolin in 2, 5 and 8 M sodium hydroxide solutions showed that higher alkalinity resulted in higher amounts of dissolved aluminium and silicon entities. The digestion rates decreased at extended times and the aluminium component in the metakaolin dissolved more rapidly than the silicon component except in the 8 M solution at extended digestion times.

NMR results confirmed the presence of $[Al(OH)_4]^-$ but no monomeric silicate was observed. This was attributed to the fast reaction of aluminates and silicates and higher dissolved concentrations of aluminate compared to silicates. The formation of a shoulder in the ^{29}Si NMR spectra at -87 ppm is attributed to these aluminosilicate products with each silicon atom co-ordinated to 4 aluminium atoms. NMR also confirmed the decrease in free water.

The condensation reactions take place in an optimum band of alkalinity. At low molarity the dissolution rate is too slow to generate sufficient reactant products for condensation to occur. If the alkalinity is too high then most of the silicate exists as $[SiO_2(OH)_2]^{2-}$ which is not favoured for the condensation process. The condensation reactions are only occurring in the range 5–10 M as confirmed by the calorimetry results.

De Siva (2008) investigated the influence of aluminium and silicon concentrations on setting times and strength development of geopolymers based on metakaolin and sodium silicate. Setting times were controlled by aluminium concentration and were shown to increase with increasing $SiO_2:Al_2O_3$ ratios of the initial mixes. Increases in aluminium concentration led to low strength products which showed grain like structures in SEM images as opposed to amorphous geopolymer phases for higher silicon content geopolymers i.e. the strength characteristics depend more on the silicon content.

The authors suggested that the aluminium component of metakaolin is more readily soluble than the silicon component and in the lower silicon: aluminium systems it is possible that more monomeric aluminate ions are available for condensation. Polysialate structures are likely to be formed in this case. With increasing silicon content more silicate species are available for condensation, which can take place between silicate species resulting in oligomeric silicates. Further condensation between these oligomeric silicates and aluminates results in a three dimensional, rigid structure based on poly (sialate disiloxo) and poly (sialate-siloxo) derivatives. The rate of condensation between silicate entities is slower than that between aluminate and silicate resulting in slower setting with increasing silicon content.

Duxson (2007) presented a simplified model for the geopolymerisation process (Fig. 3.1). Dissolution of the aluminosilicate source by alkaline hydrolysis (consuming water) releases aluminate and silicate species. These then form an equilibrium mix with aluminosilicate material and any added silicate or aluminate species from the activating solutions. In concentrated solutions, as employed in geopolymer synthesis, this results in the formation of a gel as the oligomers in the aqueous phase form large networks by condensation releasing water which is consumed in the dissolution process. This water resides in the pores of the gel. Reorganisation of gel 1 to gel 2 occurs with the release of more water. Finally polymerisation via a condensation step occurs.

Steins (2012) investigated the geopolymerisation step using rheology, small angle X-ray scattering (SAXS) and electron paramagnetic resonance (EPR). Metakaolin activated with sodium, potassium or caesium hydroxides with $Al:M^+$ adjusted to 1 was used in the investigations.

Dynamic rheological measurements were carried out using a controlled stress rheometer. Applied strain was 1×10^{-4} and frequency was 1 rad s^{-1}. The geopolymer paste was initially prepared outside the rheometer with 5 min mixing carried out in a bowl with a helical ribbon impellor and then introduced to the rheometer

Fig. 3.1 Simplified geopolymerisation process model (Duxson et al. 2007)

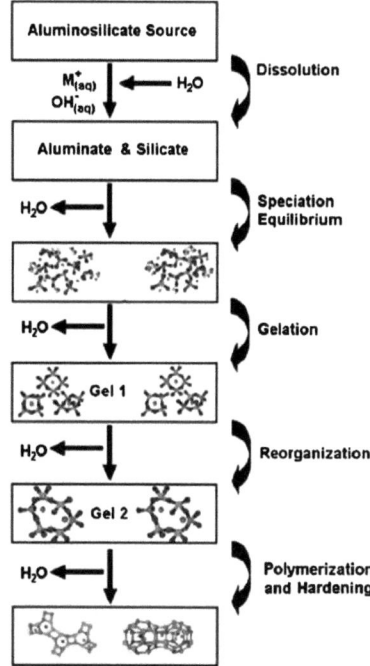

test cell. Transferring the paste to the measurement cell introduces stress which must be relieved prior to performing the test. This was achieved by applying a strain of 0.05 and a pulsation rate of 5 rad s^{-1}. Evolution of the viscoelastic parameters (G', the elastic modulus; G'', the loss modulus; tan δ, the ratio between G' and G'') was recorded during geopolymerisation at ambient temperature.

Regardless of the alkali activator used the geopolymerisation process is similar but dissolution and condensation rates are different. This was attributed to the different charge densities of the alkali cations. Small cations (Na^+) have a more compact hydration sphere than larger cations (K^+, Cs^+) which more readily bind with negatively charged silicates which are partly de-protonated at pH $\gg$ 10. The peak in the tan δ curve equates to gel time (Fig. 3.2).

SAXS traces were run on the activating solutions and metakaolin dispersed in water to set up reference points. The geopolymer reaction mixtures were then run at intervals. A decrease in SAXS intensity is characteristic of the presence of sharp interfaces between two media of different electron density. This was attributed to metakaolin particles surrounded by activating solution. As the metakaolin dissolves there is a decrease in the signal at low q values and a simultaneous increase in the signal intensity for the formation of new monomers and then the formation of secondary structural units. Finally new aggregation is observed as shown by a scattering intensity decrease that reveals a new interface. The intermediate signal due to monomers has become small suggesting conversion to oligomers is complete. Several events occur simultaneously, the dissolution of the metakaolin leading to

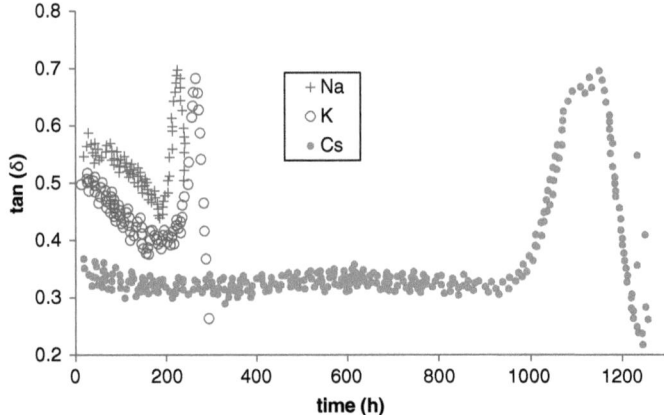

Fig. 3.2 Measurement of gel time by tan δ development using a stress rheometer at ambient temperature (Steins et al. 2012)

the appearance of monomers or oligomers in solution and the reaction of these with the activating solution leading to nano-metric structures which evolve into the geopolymer network. At the end of geopolymerisation the intensity variation indicates a sharp, new interface of the solid geopolymer separated from pore solution.

The reaction dynamics of water during geopolymerisation was followed by EPR using a paramagnetic probe. Before the formation of a percolating network water is consumed during the hydrolysis/dissolution step and then regenerated during polycondensation reactions. In each case the energy of hydration and the alkali cation hydration sphere play an important role in local arrangement, gelation and consolidation of the geopolymer network.

The paramagnetic probe is gradually masked during geopolymerisation due to protonation of the nitroxide radical and reappears during polycondensation reactions (Eq. 3.5).

$$\text{(structure with N-O•)} + H^+ \longrightarrow \text{(structure with N-OH)} \qquad (3.5)$$

De Lacaillerie (2012) presented a similar body of work to Steins. He used NMR instead of EPR. In Fig. 3.3 he identified four regions for the geopolymer mechanism.

- Region1. Initially aluminium goes into solution. Less than 1 % aluminium is in solution at any one time. This is the rate determining step for geopolymerisation.
- Region 2. Aluminium rich gel forms around the grains resulting in an elastic modulus.

3.1 Metakaolin Based Geopolymers

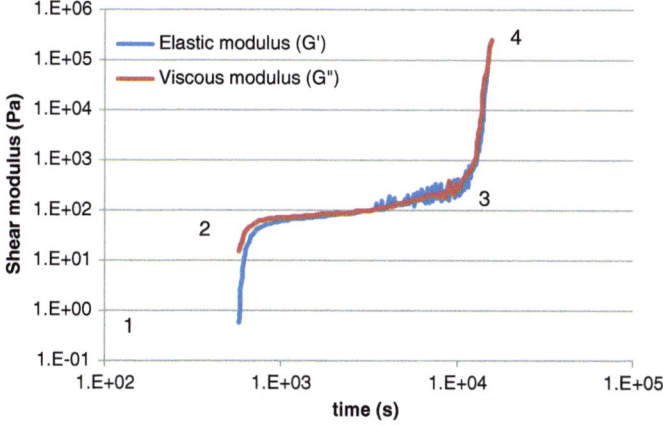

Fig. 3.3 Development of shear modulus from dissolution, condensation and geopolymerisation (De Lacaillarie et al. 2012) (Test conditions: 0.5 % strain at 1 Hz. Pre-mix 5 min at 100 Hz)

- Between regions 2 and 3 there is a transformation into a silicon rich gel.
- Region 3. Gel commences to harden and form geopolymer.
- Region 4. Cured geopolymer.

The SAXS was used to investigate the effects of $SiO_2:Na_2O$ ratio on the speciation of the silicate activating solution. NMR was used to follow changes in aluminium co-ordination in metakaolin through to hardened geopolymer.

The origin of the early elastic modulus cannot be ascribed to interactions between metakaolin particles or to the presence of sodium silicate as this is viscous. It is attributed to the formation of an aluminosilicate gel (Gel 1 in the Duxson model in Fig. 3.1).

3.2 Fly Ash Based Geopolymers

The analysis of the geopolymerisation mechanisms and kinetics are made more difficult than those of metakaolin due to the presence of significant crystalline phases (typically mullite, quartz and haematite). The alkali reactive glassy aluminosilicates may not be readily accessible for alkali dissolution reactions, being shielded by non-reactive phases. Variations between fly ash types and between different batches of the same fly ash further complicate the issue. The presence of soluble calcium entities changes the kinetics of fly ash geopolymerisation.

Fernandez-Jimenez et al. (2003) analysed five Spanish Class F fly ashes and used compressive strength to judge the suitability of the fly ashes for geopolymerisation. The major factors in fly ash composition for geopolymerisation were determined to be the reactive silica content, vitreous phase content and particle size distribution.

Fernandez-Jimenez et al. (2005) developed a model for the synthesis of geopolymers from fly ash using alkaline activation. In this model the glassy constituents of the fly ash are converted by highly alkaline activating solutions into cementitious geopolymers. Using SEM to observe changes to micro structure over the cure period at 85 °C the mechanism shown in Fig. 3.4 was developed.

Figure 3.4a shows the initial alkali attack at one point of the fly ash surface which then grows into a larger hole (Fig. 3.4b) exposing smaller spherical particles, which can be hollow or solid, to alkaline attack from the outside in and from the inside out. Reaction product is generated both inside and on the outside of the fly ash shell until the particle is almost or completely consumed (Fig. 3.4c). The mechanism involved at this stage is dissolution. One of the consequences of reaction product precipitation is that a layer of these products covers portions of the smaller fly ash spheres. This covering prevents access of the alkaline activating solution (Fig. 3.4e). As alkaline activation continues, the unreacted fly ash buried under the precipitates may not be affected by the high pH of the activating solution and a reduction in reaction rate may occur. Activation is now governed by a diffusion mechanism. Several morphologies may co-exist in the paste, e.g. unreacted particles (both glassy and crystalline), particles attacked by alkali but retaining their spherical shape, reaction product etc. (Fig. 3.4d). This model was also applicable to activation by sodium silicate solutions.

Fernandez-Jimenez et al. (2006) investigated three class F fly ashes with similar Si:Al ratios, but different reactivity, indicating that not all the silicon and aluminium entities are reactive towards alkaline activation. The degree of reaction was determined by using 5 wt% hydrochloric acid, which was purported to dissolve the reaction products of alkali activation (aluminosilicates) but not the unreacted fly ash. The concentration of dissolved aluminium was determined by ICPS-MS. Analysis of compressive strength, degree of reaction and microstructure

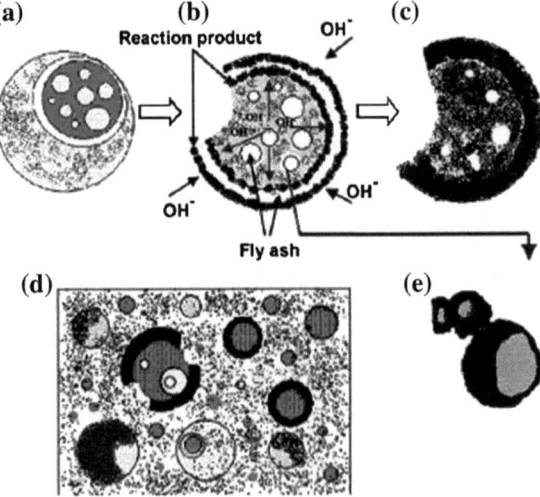

Fig. 3.4 Alkali activation of fly ash model (Fernández-Jiménez et al. 2005)

characterisation showed that the preferred fly ash for alkaline activation should have high contents of reactive silica and alumina and (Si:Al)$_{Reactive}$ of less than 2. For short reaction times gels have an aluminium rich phase in which silicon tetrahedra are surrounded by four alumina tetrahedra ($Q_4(4Al)$). As the reaction progresses this phase evolves into a more stable silicon rich phase in which higher amounts of silicon occupy $Q_4(3Al)$ and ($Q_4(2Al)$). Improvements in mechanical properties are observed as the reaction progresses. The use of this analytical method by Fernandez-Jimenez (Fernández-Jiménez et al. 2006) is contrary to the reported chemical resistance of geopolymers to hydrochloric acid suggesting further investigation of the technique is required (Chaudhary and Liu 2009; Rostami and Brendley 2003).

Van Jaarsveld and van Deventer (1999) concluded that the alkali metal cations control and affect all stages of the geopolymerisation reaction. During the dissolution process they arrange ions and soluble species and play a structure directing role during gel hardening and eventual crystallisation. Dissolution of fly ashes from different sources will dissolve with different silicon: aluminium ratios depending on the alkali cation present.

Criado (2007) investigated the effect of SiO_2:Na_2O ratio on the alkali activation of fly ash. Four activating solutions with different soluble silica contents were used. The main reaction product was a sodium aluminosilicate gel with different zeolites appearing as minor phases. The amount and composition of the reaction products depended on the soluble silica content in the activating solutions and the thermal curing conditions. The amount of gel was related to the development of mechanical strength in the geopolymer.

XRF and QXRD were used to calculate the amorphous and crystalline components of the fly ash and reaction products. Curing was carried out at 85 °C for times ranging up to 180 days. Determination of produced reaction products was carried out by dissolution in 1.76 % hydrochloric acid solution. The acid dissolves the reaction products (sodium aluminosilicate and zeolites) leaving insoluble fly ash residues behind, but see above.

At short cure times (8 h) an increase in soluble silica concentration developed compressive strength >30 MPa compared to ~15 MPa for the lower soluble silica content. After 20 h all systems were showing >40 MPa compressive strength. The long cure times (180 days) showed compressive strengths of >70 MPa across all the systems investigated.

Crystalline zeolites, sodalite and chabazite-Na were found in the low soluble silica content systems even at the short cure times. Increasing the soluble silica level retarded the formation of zeolite species to longer cure times and promoted the formation of Zeolite Y and Zeolite P. The percentage of zeolites always grows at the expense of aluminosilicate gel which may confirm that the geopolymer gel is a zeolite precursor.

The amount of vitreous phase (from the fly ash) decreased with increasing cure time in all systems. The systems with high soluble silica contents showed a slower rate of vitreous phase consumption. This was attributed to the soluble silica being highly polymerised, which reduces the rate of ash dissolution. The delay in the initial dissolution of fly ash seen in systems with high soluble silica and reduction in the activation kinetics is offset by the formation of larger molecular species with a denser more compact gel.

The quartz and mullite phases from the fly ash were observed to decline slightly at the longer reaction times suggesting partial attack by the activating solutions.

Fernandez-Jimenez (2005) first used the term N-A-S-H (sodium aluminosilicate hydrate) for the sodium aluminosilicate formed as the main reaction product of alkali activation of fly ash. Whenever an alkaline activating solution is used with fly ash an alkali metal aluminosilicate gel is formed as the main product. This gel is responsible for the mechanical properties. Alkaline activators based on sodium hydroxide, sodium hydroxide and sodium silicate and sodium hydroxide and sodium carbonate were evaluated with a class F fly ash.

The Na_2O content plays an important role in the development of physical properties. Increasing the Na_2O content in the sodium hydroxide activator system led to increases in compressive strength up to 70.4 MPa. When soluble silica (sodium silicate) was added to the activating solution compressive strength showed further increases to 90 MPa after 20 h curing at 85 °C. In addition to the $SiO_2:Na_2O$ ratio, water to binder ratio needs to be considered. Higher water contents reduce the pH (lower HO^- concentration) and reduce the rate of dissolution. The presence of carbonate ions in the activating solution leads to low strength (35 MPa) geopolymers even if Na_2O levels are high. Small amounts of zeolite crystals were detected and their presence probably indicates that these crystalline products are the thermodynamically stable phases towards which the system could evolve over time.

The mechanism controlling the chemical reaction giving rise to the aluminosilicate gel is initially associated to a dissolution process (the high concentration of OH^- ions in the system is responsible of the breakdown of the Si–O–Si, Si–O–Al and Al–O–Al bonds forming part of the vitreous phase of the ash and leading to the formation of Si–OH and Al–OH groups). Later on these chemical species condense giving place to the precipitation of precursor.

The addition of sodium silicate to the activating solution enhances the polymerisation process of the ionic species present in the system. Activating solutions made from sodium hydroxide and sodium silicate needs to be optimised in terms of not only the $SiO_2:Na_2O$ ratio but also the actual amounts. A threshold exists at Na_2O >7 % and SiO_2 >1 % above or below which mechanical strength development at 85 °C is less than 65 MPa. Equilibrium between NaOH and Sodium silicate in the solution should be reached in order to maintain the system with a high pH and a high level of soluble silica.

When the activating solution is a mixture of sodium hydroxide and sodium carbonate, the main modification induced to the system is the incorporation of carbonate to the mixture, which promotes the formation of sodium bicarbonate (trona) among the reaction products. This involves a reduction in pH leading to lower amounts of aluminium and silicon dissolved from the fly ash. This could explain the porous microstructure and low strengths obtained with this activator solution.

The alkali metal cations play a charge balancing role where aluminium ions (Al^{3+}) have replaced silicon ions (Si^{4+}).

3.3 The Role of Calcium Compounds

Van Deventer et al. (2007) found that the amount of Ca^{2+} and the form in which it is added both play a role in determining the properties of the final geopolymer. The level of dissolved silicate in the activating solution also plays a significant role in determining the effects of calcium on the final reaction product by influencing the pH of the reacting system and affecting the stability of the different calcium precipitates. The addition of highly alkaline activating solutions to a fly ash containing any calcium leads to rapid dissolution of the calcium from the ash followed by precipitation of calcium hydroxide. This has the effect of removing OH^- ions with consequent pH reduction which will influence the dissolution and condensation geopolymerisation processes. The addition of small amounts of calcium in the form of a soluble salt to a class F fly ash based geopolymer reaction system drastically increased the solidification rate and early yield stress.

Yip (2008) investigated the effect of different calcium silicate sources on geopolymerisation. Calcium dissolution from manufactured sources (Portland cement or blast furnace slag) at low alkalinity forms CSH phases in conjunction with the geopolymer gel. Less calcium dissolves from natural calcium silicate minerals with little formation of CSH phases. The undissolved mineral disrupts the geopolymer resulting in lower overall strength. The co-existence of CSH and geopolymer phases give rise to acceptable levels of mechanical properties of the binder systems synthesised at low alkalinities. At high alkalinity calcium plays a lesser role as it forms precipitates rather than hydrated gels.

Temuujin (2009b) studied the addition of calcium oxide and hydroxide at 1, 2, and 3 wt% to a fly ash based geopolymer synthesis. Curing was carried out at 20 and 70 °C. Geopolymers made with calcium hydroxide showed higher compressive strengths than those produced with calcium oxide when cured at 20 and 70 °C. Higher levels of compressive strength were obtained at 70 °C in every case.

Dombrowski (2007) added up to 20 wt% calcium hydroxide to a fly ash synthesis mix by replacing fly ash with an equivalent weight of calcium hydroxide. 8 wt% substitution of fly ash gave the best compressive strength results after curing at 40 °C.

Diaz (2010) investigated factors involved in the suitability of fly ashes for the manufacture of geopolymers. He used five fly ashes with calcium oxide levels ranging from 1.97 to 22.45 wt%. XRD scans (using a copper source) were performed on fly ashes and geopolymers. The glass diffraction maximum (GDM) is the highest point in the broad hump area. The location of the GDM in fly ash containing up to 20 wt% CaO is typical of a siliceous glass structure with 2θ values of 22.7°–27.5°. Above 20 wt% CaO the GDM remains around 2θ values of

Fig. 3.5 Changes in setting time and compressive strength with wt% CaO (Diaz et al. 2010)

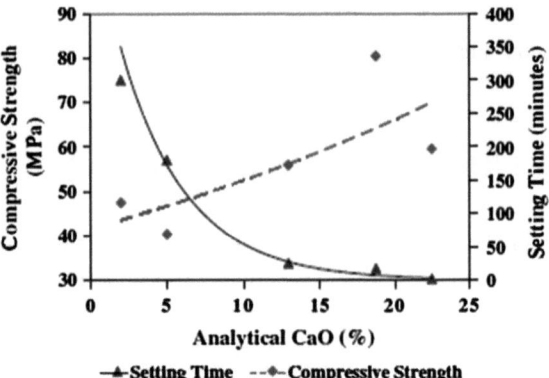

32.2°, a value typical of calcium aluminate glass structure that is generally more reactive with water compared to the siliceous glass structure. This can lead to the formation of calcium silicate hydrate structures additional to the geopolymerisation products leading to improvements in mechanical properties.

Setting times as short as 1.5 min for the highest CaO content were recorded, whilst the lowest content gave 300 min setting time. Compressive strength increased with increasing CaO content of the fly ash (Fig. 3.5).

Chapter 4
Fibres: Technical Benefits

Abstract The benefits of metallic, organic and inorganic fibres and fillers in both OPC and geopolymers are discussed at length. Cementitious materials are typically characterised by low tensile strength and strain capacity and are sensitive to micro cracking. Fibres, and/or steel and Fibre Reinforced Plastic (FRP) rebar may be incorporated into cementitious matrices to overcome these weaknesses giving materials with increased tensile strength, ductility, toughness and increased durability. The mechanism of fibre reinforcement is common to OPC and geopolymers and as such the literature covering OPC-fibre composites is relevant. The mechanism of fibre reinforcement is discussed together with comments about the effects of fibres on processibility. Fibres also contribute to improvements in durability of cementitious composites such as corrosion and fire resistance. The properties and attributes of each fibre type are outlined with respect to the result achieved in the cementitious matrix. Fibres can reduce plastic cracking in fresh concrete and improve the post crack ductility of hardened concrete. An extensive range of available fibres is covered; natural and synthetic, inorganic and organic as well as a section on carbon based fibres. Hybrid fibre blends, typically steel and polypropylene, can give synergistic effects.

Keywords Fibre reinforcement · Steel fibres · Polypropylene fibres · Polyvinyl alcohol fibres · Carbon fibres · Glass and basalt fibres · Hybrid fibre blends

Cementitious materials, typically concrete, are the most widely used materials for infrastructure construction. They are typically characterised by low tensile strength and strain capacity and are sensitive to micro cracking. Fibres and steel and FRP rebar may be incorporated into cementitious matrices to overcome these weaknesses giving materials with increased tensile strength, ductility, toughness and increased durability (ACI E2-00 2006). The mechanism of fibre reinforcement is common to OPC and geopolymers and as such the literature covering OPC-fibre composites is relevant.

The use of fibres to enhance the properties of construction materials dates back thousands of years to the use of straw and reeds in bricks and animal hair in plaster. Fibres can reduce plastic cracking in fresh concrete and improve the post crack ductility of hardened concrete. While the random orientation of the fibres means they are not as efficient as conventional reinforcement for dealing with predictable stresses, they are able to resist crack propagation under unforeseen stresses, particularly those arising close to the surface of elements during construction and service, particularly impact (Wimpenny et al. 2009).

In her Ph.D. thesis Jansson (2008) believed that the effect of fibres can be distinguished at two levels: the micro level and the macro level. The micro level covers a short stage after the linear elastic stage is surpassed, where small cracks arise from initial flaws in the matrix. As load increases the length of the microcracks increases and they coalesce and finally localise into macro-cracks. For a given fibre content, micro-fibres, due to their vast number are more likely to cross these micro-cracks. For microfibres to be effective they should have a high aspect ratio and stiffness to enable them to restrain the micro-cracks as they propagate into the cementitious binder. If an improvement in structural properties is desired e.g. in bending, then fibres must be selected with sufficient length to bridge macro-cracks and specific mechanical properties.

Jansson referred to Naaman (2003) who claimed that in order to be effective in concrete matrices, fibres must have the following properties:

- Tensile strength of approximately two to three orders of magnitude higher than that of concrete.
- Bond strength with the concrete matrix of the same order or higher than the tensile strength of the matrix.
- Elastic modulus in tension at least three times higher than that of the concrete matrix.

In order to overcome detrimental de-bonding the Poisson's ratio and the coefficient of thermal expansion should preferably be of the same order for both fibre and matrix. Introducing surface deformation to create mechanical anchorage helps increase bonding performance.

As concrete is loaded, cracks form. Initially the cracks are short, discontinuous micro-cracks, which coalesce to form large macro-cracks. Fibres bridge cracks, transferring the load and delaying the coalescence of cracks. Crack formation and development is affected by the shape, size, type and volume of the fibre reinforcement. Steel bars are the materials of choice for structural applications, but the addition of "short fibres" now plays an important role in the processing and the development of early stage properties such as plastic and drying shrinkage (Banthia and Gupta 2006). Additionally non-metallic structural fibres are finding application (Brugge 2011; Forta Corporation 1999).

The following attributes are required for a fibre to reinforce cement mortar (Karbhari 1998):

- Strain to failure higher than that of cement mortar
- Small fibre diameter

- Hydrophilic surface that give good dispersion and bonding
- Long term durability in alkaline environments
- High strength and modulus
- Overall durability to harsh external environments.

Drying and plastic shrinkage cracks are surface cracks that occur due to water evaporating too quickly from the surface of concrete during the curing process. This causes the surface of the concrete to dry quicker than the layers below and "shrink", leaving behind thousands of tiny cracks in the concrete (Cement Concrete and Aggregates 2005).

The primary function of these microfibers is to modify the properties of fresh, plastic concrete. They can improve homogeneity, reduce bleeding and reduce plastic settlement and plastic shrink cracking. The influence of microfibers on the compressive properties of hardened concrete is relatively small but they can reduce permeability and increase resistance to impact, abrasion and shatter and can reduce spalling in the event of a fire. They also provide some resistance to damage caused by frost (CCANZ 2009).

Fibres may be classified by strength/stiffness i.e. high modulus such as steel and carbon fibres and low modulus such as polypropylene. The interface between the fibres and concrete is important to determining the properties of the composite (Naaman 2008).

In OPC there is often nucleation and growth of calcium hydroxide on the surface of the fibres which results in the formation of a porous layer about 10 μm from the surface of the fibre (Bentur et al. 1985). This interface between fibre and matrix means an approaching crack can be diverted along the interface and run parallel to the fibre. The failure strain is only a fraction of the fibre yield strain and fibre pull out occurs rather than tensile failure of the fibre.

In High Performance Concrete (HPC) the presence of microsilica and low water/cement ratio used means that the porous layer is eliminated and fibre anchorage improved. This leads to improvements in strength, deformability, impact resistance and drying shrinkage of HPC by fibre additions (Gani 1997). The addition of micrometre sized diameter fibres is critical to the survival of HPC in the event of high temperature excursions such as fire exposure (Papworth 2000).

Fibres can be beneficial under extreme environments, such as chloride exposure and fire situations. Fire and wear resistance are enhanced and the discrete nature of fibres means that the risk of corrosion and spalling are reduced (Greenhalgh 2003; Perry 2006).

The mining and tunnelling industry make extensive use of fibres in sprayed concrete linings. Here fibres allow the lining to retain ductility under high deformation which is critical for safety. Precast tunnel segments have utilised both steel and organic fibres for component handling and fire resistance respectively. Marine installations use macro organic fibres to eliminate corrosion risk due to sea water exposure (Brugge 2011; Wimpenny et al. 2009).

There are applications such as curtain walls and ducting where cement mortar could be used if tensile strength, flexural modulus and toughness could be

Fig. 4.1 Effect of fibres on the tensile performance of cementitious composites (Kuder and Shah 2010)

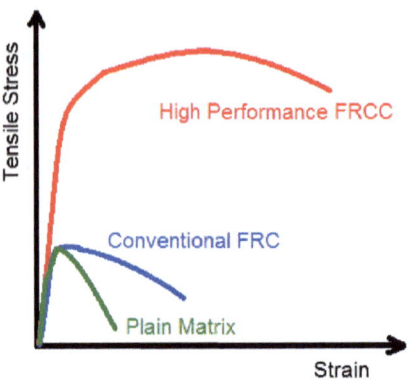

improved. Asbestos fibres in the chopped form had been used for reinforcement in thin boards and panels to increase toughness and prevent cracking. Potential replacements for asbestos have ranged from steel fibres, polyolefin and nylon fibres through to aramid, glass and carbon fibres (Karbhari 1998).

Kudar and Shah (2010) investigated the manufacture of cellulose fibre-cement boards with improvements in mechanical strength and freeze-thaw resistance to products made from the Hatschek process over those from extrusion processes. Figure 4.1 demonstrates that unreinforced cementitious materials show strain softening (plain matrix curve) with low tensile strength and ductility. Conventional fibre reinforced composites (FRC) with typically 0.5–2 vol% fibre also show strain softening with an increase in post peak ductility. High performance fibre reinforced cementitious composites FRCC (30–35 vol% fibre) show an increase in elastic limit, followed by a strain hardening response as multiple cracks form, but do not widen and finally strain softening as cracks widen.

4.1 Reinforcement

The use of fibres in OPC and Geopolymers is analogous to polymer (FRP, GRP) based composites which are in wide spread use.

The reinforcing fibres can be based on polymers, metals or ceramics and supplied in several forms such as:

- Rigid bars (Rebar)
- Woven mats
- Continuous cords and rovings
- Chopped fibres
- Nonwoven mats
- Milled fibres
- Nano material
- Whiskers.

4.1 Reinforcement

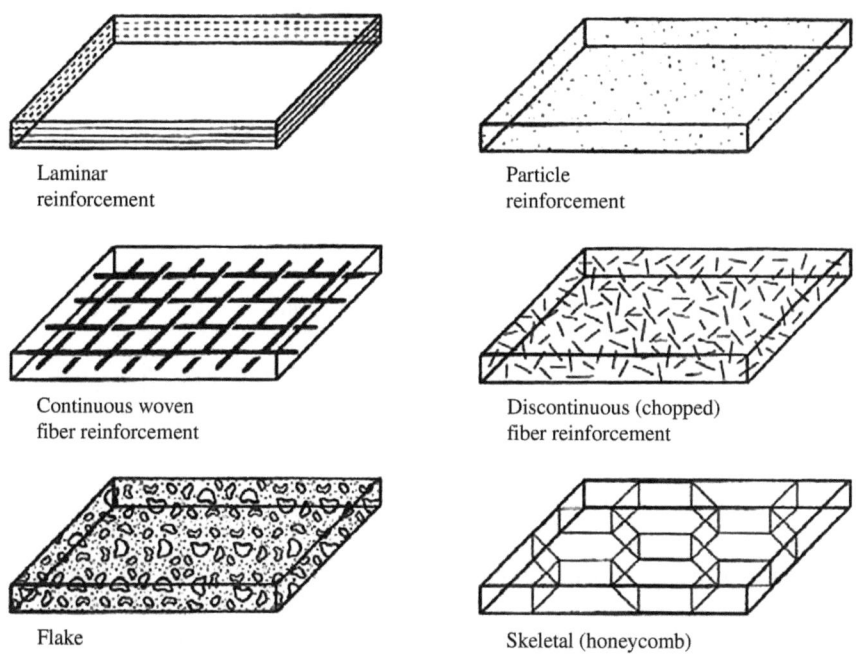

Fig. 4.2 Common forms of glass fibre for composite reinforcement (Budinski and Budinski 2005)

Conventional FRP composites contain from 20 to 50 wt% of glass or other reinforcement. In advanced composites based on epoxy resins and graphite this figure can increase to 70 wt%. Figure 4.2 shows some commonly used reinforcement forms. Chopped fibres, flakes, particles and similar discontinuous reinforcement forms are usually not as effective as continuous reinforcements in increasing creep strength. However, discontinuous reinforcements enable less labour intensive application due to direct addition to the liquid matrix phase (Budinski and Budinski 2005).

4.2 Steel Fibre Reinforced Concrete (SFRC)

Katzer (2006) reviewed the history of steel fibre reinforced concrete. In 1874 Bernard, in California patented the idea of strengthening concrete with the addition of steel splinters. In 1910 Porter added short wire to concrete already reinforced with thick wire. In 1918, in France, Alfven patented a method of modifying concrete by long steel fibres in order to increase the tensile strength of concrete. He was the first to mention the effect of surface roughness on their adhesiveness to the matrix and the problems of anchorage of the fibres. Steel fibres

Fig. 4.3 Steel fibre types for concrete reinforcement (The Concrete Society 2007)

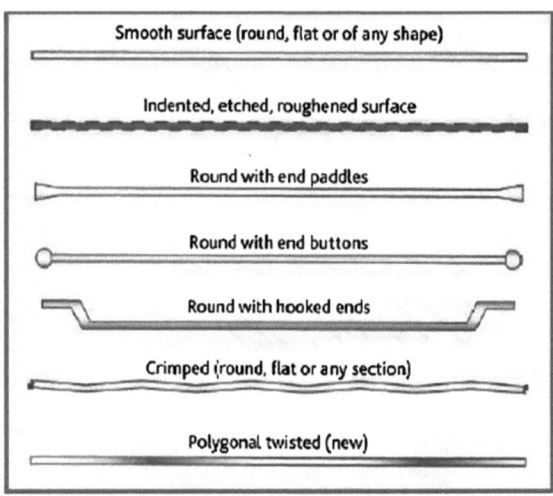

patented by Constantinescu in 1943 were similar to existing fibres but gave information about the type and dispersion of cracks during loading of SFRC elements. It also referenced the large amount of energy absorbed by SFRC under impact.

Over 90 % of produced steel fibres are shaped; the shapes adjusted to increase the anchorage of fibres in concrete. Length and cross section area can be varied for effect. Figure 4.3 shows the commonly available steel fibre types. 67 % of fibres produced consist of a hook type.

The efficiency of dispersing fibres in concrete depends mainly on the aspect ratio, which effects workability and spacing throughout the concrete. To maintain workability aspect ratio should be less than 150. 50 % of all fibres produced worldwide lie in the range 45–63.5 (Katzer 2006).

ASTM A820/A820M-11 (2011) has five types of steel fibre classified by the process or product used as a source for the steel fibres.

Romualdi et al. (Romualdi and Batson 1963; Romualdi and Mandel 1964; Swamy and Mangat 1974) were amongst the first to clarify the science behind the functioning of steel fibres in concrete. Initial work in the early 1960s brought the use of steel fibres in concrete to the greater awareness of industry and pioneered the growth of SFRC applications.

The first major application of SFRC was to build airport runways in the USA. 28 installations were constructed between 1972 and 1980 using various steel fibre types at addition levels of 0.3–2.0 vol%. Ongoing inspections showed only scarce cracks and local damage. Following on from this SFRC was used to build motorways, dams and canals. Shotcrete SFRC has been used to stabilise embankments and landslides in conjunction with previously installed steel mesh. Experience with this system built up confidence in shotcrete SFRC to the point where the labour intensive steel mesh installation could be eliminated (Katzer 2006).

The addition of steel fibres to precast components significantly reduces the risk of cracks during transport and assembly. Tunnel linings have been installed in this

4.2 Steel Fibre Reinforced Concrete (SFRC)

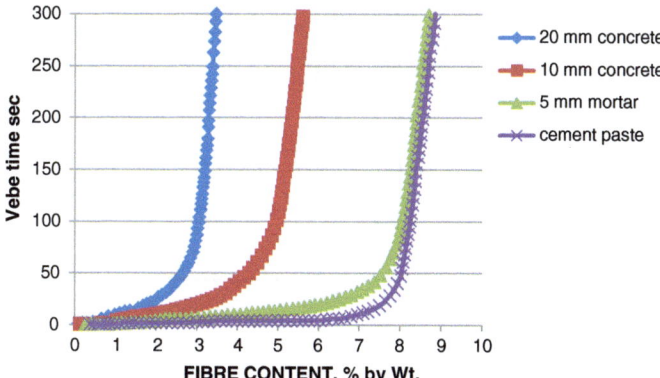

Fig. 4.4 Workability versus fibre content for different aggregate sizes (Chanh 2004)

fashion. Spun cast pipes up to 13 m long are made in Sweden using SFRC which imparts improved durability to the end product. Prefabricated insulating panels with SFRC skins and a thermally insulating core are produced in Poland.

Chanh (2004) reported that SFRC composites showed superior resistance to cracking and crack propagation. These composites possessed increased extensibility and tensile strength, both at first crack and ultimate, particularly under flexural loading. The fibres were able to hold the composite together. Figure 4.4 shows that workability decreases as aggregate size increases above 5 mm, whilst below 5 mm sized aggregate there is little influence. Workability also decreases with increasing fibre length and it is difficult to obtain good workability above an aspect ratio of 100.

One of the main difficulties in processing SFRC is the tendency for steel fibres to clump or ball together. Clumping may be caused by a number of factors:

- The fibres already clumped prior to adding to the mix. Normal mixing action will not break down these clumps.
- Fibres added too quickly to allow proper dispersion in the mix.
- Too high a volume of fibres may be added.
- The mixer is too worn or inefficient to achieve correct dispersion.
- Adding fibres before other ingredients may cause clumping.

The use of collated fibres held together with a water soluble size which dissolves in the mix water largely solves the clumping problem.

Ross (2009, 2012) states that combining SFRC with steel mesh enables the engineer to design for acceptable crack control that can reduce or eliminate joints. This would be beneficial in liquid tight concrete such as secondary bunds, water tight basements and dangerous goods stores. Using steel fibres can lead to a 50 % reduction in conventional reinforcing with resulting material and labour savings.

Hameed et al. (2009) compared two stainless steel fibres with differing aspect ratios (125 vs. 105) in OPC. Flexural toughness with 2 vol% added fibre increased

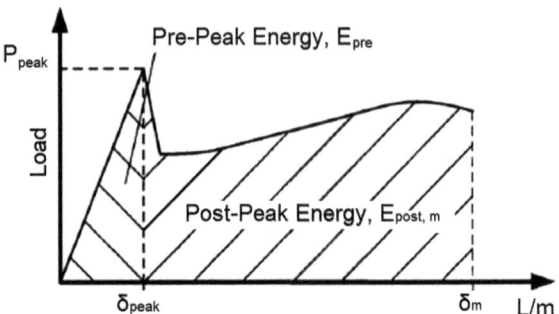

Fig. 4.5 PCS analysis on an FRC beam (Hameed et al. 2009)

approximately threefold for the fibre with the 105 aspect ratio and approximately fivefold for the 125 aspect ratio compared to a control with no fibre. He also reported a Post Crack Strength (PCS) method for converting load deflection curves into equivalent flexural strength curves (Fig. 4.5).

Stainless steel fibres are available for refractory applications beyond 1,200 °C where conventional steel fibres are unsuitable (Fibretech 2001).

Wang et al. (2010) found that the three point bending strength and toughness of the steel reinforced mortar are related to the interfacial characteristics and microstructure morphology near the fibre surface.

Bernal et al. (2006) found that alkali activated slag concretes reinforced with steel fibres exhibited higher mechanical properties than an OPC control with respect to toughness in both un-notched and notched samples.

Eswari (2008) reported on the use of blends of different fibres in concrete. The use of two or more fibres can produce a composite with better engineering properties than the individual fibres. These synergistic blends of fibres are commonly referred to as hybrid fibre blends. Typical blends are steel and polypropylene fibres, where the steel fibre improves the first crack and ultimate strengths and the polypropylene fibres imparts improved toughness and strain capacity in the post cracking region. The fibres used in this work were a macro polypropylene fibre (length = 48 mm; aspect ratio = 40) and hooked steel fires (length = 30 mm; aspect ratio = 60). They found that a 2 vol% hybrid fibre blend of 70/30 steel/polypropylene enhanced ductility and improved flexural strength.

Peng (2006) investigated hybrid blends of steel and polypropylene fibres to improve spalling resistance at temperatures up to 800 °C. He found that the hybrid blend showed better overall performance in improving spalling resistance than tests involving both fibres separately.

4.3 Organic Fibres

Several different polymer fibres are in common use (Table 4.1). They can be classified by performance and size into macro and micro fibres.

4.3 Organic Fibres

Table 4.1 Comparison of fibre properties

Fibre description	Supply form	Fibre diameter (μm)	Aspect ratio	Tensile strength	Young's modulus (GPa)	Strain to failure (%)	Density (kg/m^3)	Specific strength[a]	Temperature limit (°C)	Alkali resistance	Damage resistance	Comments
Steel fibre	Various	150–1,000	100+	0.4 GPa	200	4–10	7.86	0.25	550	Good	NED	
Glass E glass	Monofilament	15	900	3.5 GPa	76	4	2.6	1.33	600	Poor	Poor	
AR glass (zirconia based)	Monofilament	13	900	2.5 GPa	70	1–3.5	2.7	0.96	600	Good	Poor	16 % Zr minimum
Polyethylene (PE)	Fibrillated/monofilament	25–1,000	50	0.2 GPa	5	8–10	0.95	0.22	110	Excellent	NED	Low melting point
Polypropylene (PP)	Fibrillated/monofilament	20–1,000	50	0.5 GPa	4	5–25	0.91	0.6	150	Excellent	NED	
Polyamide (nylon)	Fibrillated/monofilament	20–400	50	1.0 GPa	5.6	16–20	1.14	0.87	200	Good	NED	Nylon 6.6
Polyvinyl alcohol (PVOH)	Fibrillated/monofilament	27–660	50	1.3 GPa	30	8–11	1.3	1	175	Excellent	NED	
ARAMID (Kevlar)	Pulp	10		2.8 GPa	58	4	1.44	1.92	480	Poor	NED	Kevlar 29
Wood cellulose	Pulp	25–75		0.7 GPa	4.8	8	1.5	0.47	200	Poor	NED	
Carbon fibre PAN based	Monofilament	7–10	10–1,000	4.3 GPa	225	1.90	1.82	2.36	400	Excellent		Tenax chopped fibre
Pitch based		18		1.3 GPa	31	2–2.5	1.65	0.76	400	Excellent		Isotropic pitch based
Carbon nanofibres (CNF)		0.06–0.2	500–1,500	7 GPa	600		1.8	3.89	400			Pyrograf III (PR-24-PS)
Silicon carbide		10		2.7 GPa	310		2.95	0.92	1,200	Good		Sylramic, COI ceramics

(continued)

Table 4.1 (continued)

Fibre description	Supply form	Fibre diameter (μm)	Aspect ratio	Tensile strength	Young's modulus (GPa)	Strain to failure (%)	Density (kg/m^3)	Specific strength[a]	Temperature limit (°C)	Alkali resistance	Damage resistance	Comments
Alumina		10–12		2 GPa	245		3.9	0.51	1,000	Good		Nexel 720 ex 3M
Basalt	Monofilament	7–15		4.3 GPa	110	3.30	2.65	1.62	700	Good		BCF, fibres unlimited
Asbestos		0.02–0.05		1 GPa	110	1–2	3.2	0.31	900	Excellent		Comparison
Concrete, OPC				5 MPa	30	0.02	2.7	0.002	350			Comparison
Cement matrix				3.7 MPa	1–5	0.02	2.5	0.0015	350			Comparison

[a]Tensile strength/density; *NED* not easily damaged

4.3 Organic Fibres

Polymeric macrofibres were introduced in the 1990s. They are used to control cracking due to shrinkage and thermal movements and to increase post cracking energy absorption (toughness) (CCANZ 2009).

Most macro fibres have dimensions similar to steel fibres, but are made from polymers with a density in the range of 0.9–1.2 g/cm^3. Typical diameters are in the range of 0.5–1.0 mm, with tensile strengths of 350–700 MPa and typical modulus of elasticity values of 3–10 GPa. The shape of the fibres can range from cylindrical to rectangular with a crimped or ribbed surface, whilst others are thin and flat. They are generally from 40 to 60 mm long with aspect ratios from 70 to 90 (CCANZ 2009).

Macrofibres rely on an adequate level of bonding to the cement paste. Flat shaped fibres are designed to increase surface area to volume ratio. Macrofibres are beneficial when crack widths greater than 0.5 mm can be tolerated. This is typically in applications such as shotcrete for ground support applications and concrete slab on grade (CCANZ 2009).

Microfibres (CCANZ 2009) are specially designed for concrete and their primary function is to modify the plastic properties of concrete. They increase the homogeneity, and reduce bleeding, plastic settlement and plastic shrinkage cracking.

The effect of microfibres on hardened concrete properties is limited, but they can reduce permeability, increase resistance to impact, abrasion and scatter. Freeze-thaw and spalling resistance may also be improved (CCANZ 2009).

Characteristic dimensions of microfibers are 5–30 mm in length and diameters of a few tens of micrometres. Most commonly available in polypropylene they are also available in polyamide, polyester, acrylic and polyvinyl alcohol. (Note: ceramic fibre based on glass and basalt is available in this format). They are also classified as monofilament or fibrillated.

Microfibres function by evenly distributing tens of millions of fibres throughout the volume of concrete in every direction to control cracking of the concrete in the plastic state. The fibres intersect micro-cracking that occurs as the concrete shrinks. The fibres provide enough strength to prevent the micro-cracks from widening during the first few hours after the concrete is placed. Fibrillated fibres offer longer term crack resistance than monofilament types due to their structure and greater physical dimensions (CCANZ 2009).

Monofilament microfibres improve spalling resistance in fire situations by vaporising and hence forming channels to allow water escape (Adfil 2010).

Microfibres do not provide any appreciable amounts of residual strength in concrete after cracking.

The levels of fibrillated fibres used are slightly higher than that used for monofilament types, as they are thicker, to enable an equal number of fibres to be added. Typically 0.25 % by volume is added.

4.3.1 Polypropylene and Other Polyolefin Fibres

The drive to replace asbestos fibres for health and safety reasons in the 1960s led to the introduction of polypropylene, amongst other materials, for concrete reinforcement.

ASTM D7508M-10 (2010) gives a good description of polyolefin fibres. ASTM C1116-10a classifies polyolefin fibres as Type 111 (2010).

Polypropylene fibres may be used in different ways to reinforce cementitious binders. In thin sheet like components it supplies primary reinforcement but must be present at high (>5 vol%) fractions. A second application is as a secondary reinforcement where low modulus polypropylene fibres at around 0.1 % by volume are added to reduce plastic shrinkage with no influence on cured composite crack control (Banthia and Gupta 2006).

These low modulus polypropylene fibres are also used for fire protection. In the event of fire the fibres melt/vaporise leaving channels to facilitate the movement of water and reduce the tendency for spalling (Adfil 2010; Kalifa et al. 2001; Peng et al. 2006).

Wang and Niu (2012) found that 0.1 vol% of polypropylene micro-fibre improved freeze thaw resistance of OPC when de-icing salt was used.

Mazeheripour et al. (2011) evaluated the effect of polypropylene microfibres on the production of self-compacting concrete. High levels (0.35 vol%) of fibre impacted adversely on the rheology of the mixes even when additional superplasticiser was added. At 0.15 vol% addition of fibre adequate flow was achieved and improvements in splitting tensile strength and flexural strength were recorded. This paper gave a detailed account of mix design to conform to the requirements for self-compacting concrete outlined by Japanese Society of Civil Engineering (JCSE).

Toutanji et al. (1998) compared permeability and impact resistance in OPC using two lengths (12.5 and 19 mm) of polypropylene fibre at a range of loadings up to 0.5 vol%. Silica fume (5 and 10 wt%) was included in the investigation. Longer fibres showed increased permeability, but silica fume at both loadings showed marked reductions in permeability. The longer fibres showed marginally higher impact strength, but again silica fume showed much higher results. The effect of the silica fume is to improve dispersion of the fibres in the matrix. Nili and Afroughsabet (2010) carried out a similar evaluation and confirmed Toutanji's findings.

Puertas et al. (2003) compared polypropylene micro-fibre in alkali activated slag and fly ash based geopolymer, a blend of these two and OPC. The fibres improved impact strength, but other results did not show any advantage for the fibres.

Zhang et al. (2009) investigated the effect of 3 mm long × 10 μm diameter fibres with a density of 1.38–1.40 g/cm^3 and an elastic modulus of 13.5 GPa in a metakaolin-fly ash geopolymer. Increasing the fibre content from 0 to 0.75 wt% reduced the fluidity but had little effect on setting time. Flexural strength and impact strength were improved by fibre additions. He states that geopolymer shrinks during curing leading to the formation of micro-cracks and even some visible cracks and reported reductions in shrinkage of up to 44.6 % at 0.75 wt% fibre addition. SEM showed fibres bridging over pores.

Akers et al. (2009) described a bi-component polyolefin macro-fibre. This had a high modulus core (10 GPa) with a sheath material which included an embossed surface optimised for adhesion to the cementitious matrix. He detailed a test method for determination of critical fibre length and reported test results compared to steel fibres. This type of product is available commercially for use in concrete and shotcrete (Brugge 2011).

Forta Corporation (1999) employed a different approach to macro-fibre construction. They took a twisted bundle of non-fibrillating monofilament and combined it with a fibrillating fibre system. This synergistic combination is 54 mm long and can perform the functions of a micro and macro fibre. Tensile strength of the fibre is reported as 620–758 MPa, but no elastic modulus value was given. The addition of 2 vol% of these fibres enhanced residual strength after first crack.

Hu et al. (2010) produced light weight concrete containing a 38 mm long × 0.5 mm diameter polypropylene fibre. The modulus of the fibre was reported as 4.2 GPa and the density as 0.91–0.93 g/cm^3. The slump of all the mixes was controlled at 210–230 mm. Flexural strength, splitting tensile strength and toughness were all improved by addition of the polypropylene macro-fibres.

Ikai et al. (2006) developed a polypropylene fibre suitable for the manufacture of fibre-cement board using the Hatschek process. The aim was to replace expensive polyvinyl alcohol which had in its turn replaced asbestos fibres. A major part of the development was activation of the fibre surface to increase adhesion to the cementitious matrix. Interfacial friction bonding increased from 0.22 MPa for the initial untreated fibre to 0.7 MPa for the latest generation treated fibre. An average modulus of 9.1 GPa and an elongation of 20.9 % were achieved over a 12 month production period.

With asbestos and polyvinyl alcohol fibres the adhesion levels are very high causing fibres to break rather than "pull out" limiting the amount of energy absorbed. With the polypropylene fibres additional energy is required to debond the fibres and then pull them out of the matrix. This leads to superior impact resistance with products containing the polypropylene fibres.

A polypropylene fibre (Hansen 1994) called Krenit produced by splitting extruded tapes, which were pre-stretched and heat treated has been produced. The following properties are claimed: Tensile strength = 600 MPa, Young's modulus = 8.5–12.5 GPa, Ultimate Elongation = 5–8 %. The surface of the fibres is electrically treated to improve adhesion. The fibres are available in two cross sections, 100 μm × 20 μm and 300 μm × 30 μm and lengths from 3 to 20 mm.

Fibres based on ultra high molecular weight polyethylene (UHMWPE) are available (Honeywell 2010). They are characterised by high Young's modulus values (70–80 GPa). Wang et al. (1991) compared UHMWPE fibres, aramid fibres and conventional polypropylene fibres in OPC mortar. In samples containing well dispersed fibres (UHMWPE and PP) fibre pullout and stretching were

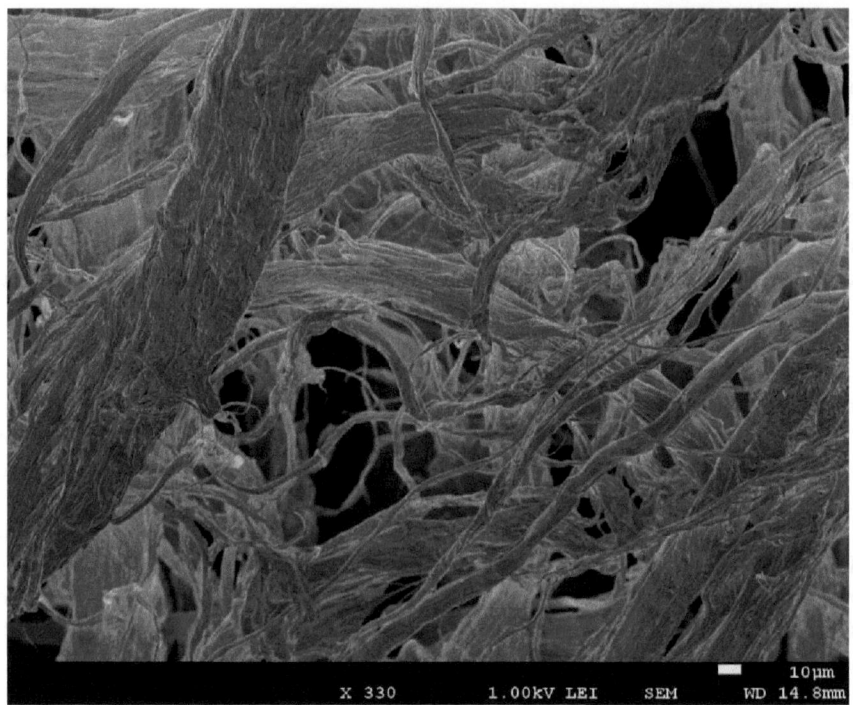

Fig. 4.6 Fibrillated polyethylene (Minifibers 2006)

the dominant mechanism. In the case of the aramid samples fibres were mainly present as bundles. These bundles cause the propagating cracks to deflect out of plane thereby increasing the splitting tensile strength. However, energy absorption was relatively low with the aramid due to lack of fibre pullout.

Low melting point grades of polyethylene are available (Minifibers 2010). Their primary function is to melt and provide release of gases with resultant reduction in pressure build up.

The unique properties of asbestos fibre made it impossible to replace with a single fibre (Minifibers 2006). In the Hatschek process a blend of reinforcing and process fibres are required to replace asbestos. The process fibre is required to trap cement particles and control drainage. Fibrillated polyethylene is able to function as a process fibre in cement fibre production. Figure 4.6 shows the extensive fibrillation of a polyethylene fibre.

4.3.2 Polyvinyl Alcohol Fibres

Polyvinyl alcohol fibres (PVOH) were one of the initial fibres considered as an asbestos replacement (Minifibers 2006; Ikai et al. 2006). PVOH fibres are characterised by high tensile strength and Young's modulus (900–1,600 MPa and 20–40 GPa, respectively). Typical elongation at break is 7–8 %. Specific gravity is 1.3. They also exhibit outstanding acid and alkali resistance (Kuraray 2007).

The chemical structure of PVOH with a pendant hydroxyl group enables very high bond strengths to be evolved with cementitious matrices. This may be a disadvantage in some applications e.g. strain hardening cementitious composites (SHCC), where fibre treatment strategies are required to give controlled fibre pull out. Oiling of the fibres is used to control fibre bonding (Kendall et al. 2008; Rotstein 2011). van Zijl and Wittmann (2010) reviewed typical SHCC formulations based on PVOH fibres and one high modulus polyethylene fibre. Increasing the oiling agent and PVOH fibre volume increased ultimate tensile strength and strain, increased first crack stress and reduced crack width and spacing. Polyethylene fibre did not require oiling.

This SHCC concept was developed by Li and Maalej (1996) and named engineered cementitious composites (ECC) (Li 1998). ECC generally exceeds 1 % strain capacity with the most ductile composite in the 6–8 % range. Coarse aggregates are not used in ECC as they affect the ductile behaviour of the composite. In general less than 2 % of discontinuous fibre is used with PVOH fibre being the most common. Fibres used in ECC must have tensile strengths of 1,000 MPa, an inelastic failure strain greater than 5 % and a fibre diameter between 30 and 50 μm (Lepech et al. 2008).

Unoiled PVOH fibres, available in micro and macro-fibre styles, can be used to produce conventional fibre reinforced concretes (Nycon 2011, 2012).

Yunsheng et al. (2006) produced extruded geopolymer boards based on metakaolin and fly ash and reinforced with PVOH fibres (6 mm long × 14 μm diameter). In the metakaolin geopolymers toughness of the boards increased as the fibre volume content increased. As the fly ash content of the geopolymer increased and with fibre kept at 2 vol%, the toughness peaked at 10 wt% fly ash addition. The fly ash greatly improved the extrusion performance of the mixes. The same authors (Yunsheng et al. 2008) reported more fully on the extrusion process. The addition of fly ash reduced the bulk yield stress and total extrusion pressure. Durability testing of the extruded boards was included with good resistance to freeze thaw and sulphuric acid (pH = 1) recorded.

4.3.3 Other Organic Fibres

Many other polymers can form fibres; polyamides, polyesters, acrylics, aramids (Zhang et al. 2011) and a host of naturally occurring cellulose based products (cotton, hemp, sisal, jute) and fibres derived from sugar, coconut and banana processing are available (Cement and Concrete Institute 2010).

Recycled fibres from carpets have also been investigated for concrete reinforcement (Nycon 2009). Nycon have commercialised a shaped (trilobal) monofilament produced from recycled nylon, which exhibits higher mechanical properties than polypropylene fibres.

The use of naturally occurring fibres for concrete reinforcement is of interest for economic and environmental reasons. Sisal fibre has been used to make roof tiles, corrugated sheets, pipes, silos and tanks. Elephant grass reinforced mortar has been used in low budget housing projects. Wood derived cellulose fibres have commercial applications in the manufacture of flat and corrugated sheeting and non-pressure pipes (Cement and Concrete Institute 2010).

Natural fibres may be used in the chemically unprocessed state or as is the case with wood cellulose processed via the Kraft process which involves digestion in alkali to reduce the lignin content. Lignin can retard or completely inhibit cement set due to its sugar content.

Davis (2007) presented an overview of natural fibre reinforced concrete. Clumping of fibres during mixing and durability were noted as two areas requiring redress.

ASTM D7537 (2012), cellulose fibres for concrete references ASTM D6942, stability of cellulose fibres in alkaline environments. The alkali resistance of natural fibres needs to be addressed for cementitious applications (Hercules Fibres 2011a; Buckeye Building Fibres 2009; Tonoli et al. 2009).

Alzeer and MacKenzie (2012) added 5 wt% of wool (washed and recycled) to a metakaolin based geopolymer and achieved up to 40 % increase in flexural strength. Only partial degradation of the wool was observed under geopolymerisation alkaline conditions. The surface of the wool was treated with formaldehyde to improve the alkali resistance.

The use of nano and micro cellulose fibres in cement mortar was investigated by Peters et al. (2010). The work showed that the addition of a hybrid blend at 3 vol% fraction increased the fracture energy by 50 %. Changes to the superplasticiser addition levels were required to maintain consistent workability, and water levels increased with fibre addition.

Okada et al. (2011) used the poor alkali resistance of polylactic acid (PLA) fibres to manufacture a porous metakaolin geopolymer to enhance capillary rise for passive cooling applications. Large amounts (13–28 vol%) of PLA fibre increasing as the fibre diameter increased from 12 to 29 μm diameter. All the fibres were 0.5 mm long. Combinations of alkali and temperature based on the melting point and thermal decomposition temperatures of the fibres were evaluated to "remove" the fibres and leave a porous structure behind. A combination of NaOH solution (pH = 12) and heating at 330 °C gave the best results with 28 % of the 29 μm diameter fibre giving the best capillary rise result.

4.3.4 Carbon Based Reinforcing Fibres

Reinforcement of cementitious materials is typically provided at the millimetre and/or micro scale using micro and macro fibres respectively. Cementitious materials also exhibit flaws at the nano level where traditional reinforcement is ineffective (Metaxa et al. 2010). The use of nanofibres such as carbon nanotubes (CNT)

4.3 Organic Fibres

and carbon nanofibres (CNF) are expected to show several advantages as a reinforcement material for cement compared to traditional fibres.

- Greater strength and stiffness compared to traditional fibres
- Higher aspect ratio is expected to arrest nano cracks and require greater energy for crack propagation
- Nanofibres are expected to be closely spaced in the cement matrix (if correctly dispersed).

Besides the benefits of reinforcement, autogenous shrinkage results indicate that CNTs can have beneficial effects on transport properties of cementitious materials leading to improved durability.

Carbon fibres are available in three basic types:

- Macro fibres, longer than 1 mm with diameters around 7 μm.
- Microfibers, with lengths in the 100 μm range and diameters up to 20 μm.
- Nanotubes and nanofibres with lengths in the 2–4 μm micron range and diameters of around 100 nm.

Carbon macro and mini fibres are produced by graphitisation of fibres (polyacrylonitrile or rayon) or of aromatic pitches based on coal tar. The latter material yields lower production costs, but gives fibres with lower stiffness properties.

Carbon nanotubes (CNT) and carbon nanofibres (CNF) are attracting a lot of research interest in OPC and geopolymers. Industrial production is limited to probably a few hundred tonnes per annum, but Bayer commissioned a 200 tpa pilot plant in Leverkusen in 2010. Bayer believes that the carbon nano material market will grow at 25 % per annum with a potential market of several thousand tonnes per annum. This will be mainly in the engineered plastics market segment (Bayer Materials Science 2007).

Dispersion of carbon nano materials into the matrix material is an ongoing issue. The de-aggregation of nano particles must be carried out in a manner that does not damage the primary reinforcing particles and interfere with the cement hydration process (Yazdanbakhsh et al. 2010). They used ultrasonication and a superplasticiser to achieve a water dispersion of CNFs in water. This dispersion was added to cement to achieve a 0.4 wt% ratio (based on cement). The dispersion of the CNFs was not uniform and this was attributed to cement particles absorbing superplasticiser and causing CNFs to re-agglomerate. One suggested solution is to functionalise the surface of the CNFs to improve wettability.

Bayer Material Science (2008) claim that carbon nanotubes can be dispersed in low viscosity media such as water using a jet disperser or ultrasonic treatment. Stabilisers need to be added to aqueous dispersions to ensure that re-agglomeration of nanotubes does not occur.

Eden Energy has undertaken research involving concrete reinforced with carbon nano-fibres (CNF) and carbon nano-tubes (CNT). The nano-carbon is produced by Hythane, a US subsidiary of Eden Energy. Hythane used a surfactant to disperse the carbon nano materials which showed increases in 7 day flexural strength (ASTM C348) ranging from 15 to 30 % (Eden Energy 2011).

Applied Sciences Inc. (2001) manufacture the Pyrograf range of carbon fibres, carbon nanofibres and microfibers. These materials are claimed to mitigate crack formation in concrete. The coefficient of thermal expansion (COTE) of these materials is listed as −1.0 ppm. This will lead to a reduction in the COTE of composites containing these materials coupled with lower thermal strain.

ASI also claim that CNF reinforced concrete is highly resistant to micro cracking, which can reduce failure of cement and concrete products due to "freeze-thaw" cycles (Ohio Coal Development Office 2004).

Cwirzen et al. (2009) used functionalised (with –COOH) multiwalled carbon nanotubes (MWCNT) and carbon micro fibres (CMF) to investigate crack formation and propagation of cement based matrices. Internal tension forces were introduced by repeated freeze-thaw cycles. The sample containing CMFs exhibited no micro-cracking and showed significant increase in flexural strength compared to the reference (no fibres) and that containing MWCNTs. The fibres were dispersed in water using a polycarboxylate superplasticiser and ultrasonic energy. A mini cone flow test was carried out, with levels of MWCNT of 0.5 wt% inhibiting flow. Up to 2.9 wt% of CMF had 90 % of the reference flow. The level of sonication affected the morphology of the nanofibres, causing length reduction which prevented efficient transfer of stress.

The Transportation Research Board of the USA recently completed a project entitled "Crack Free Concrete made with Nanofiber Reinforcement". Metaxa (2010) reported on this work. Cementitious matrices exhibit flaws at the nano-scale where traditional reinforcement is ineffective. Evidence from SEM imaging showed that CNFs were able to control cracking by bridging nano-cracks and pores in the cementitious matrix. Additionally SEM showed good bonding between the CNFs and cement hydration products indicating that the nanofibres can be sufficiently secured in the matrix to ensure the full capacity of the fibres is utilised to transfer the load (Fig. 4.7).

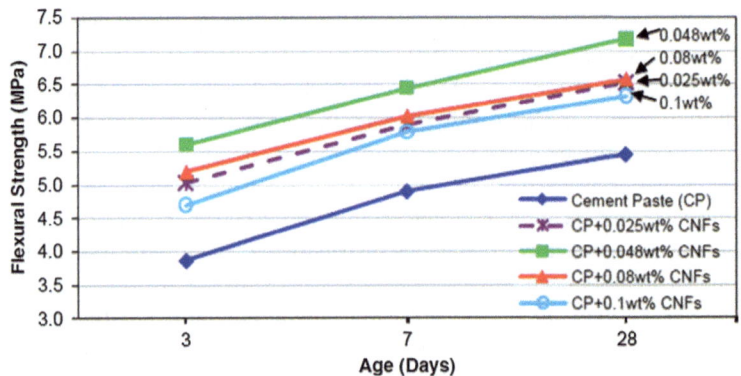

Fig. 4.7 Effect of CNF concentration on flexural strength of cement paste (water-cement ratio = 0.5) (Metaxa et al. 2010)

4.3 Organic Fibres

Chung (1992) added short pitch based carbon fibres (0.05 wt% of cement) to concrete and obtained a flexural strength increase of 85 % and a flexural toughness increase of 205 %. The air content was 6 %, so freeze-thaw durability was improved, even in the absence of an air entrainer. The fibre length decreased from 12 to 7 mm during mixing. The drying shrinkage was decreased by up to 90 %. She found that pre-dispersing the carbon fibres in water using a dispersing agent and a defoamer was the most practical approach.

Lin (2009a) reinforced a metakaolin based geopolymer (silicon to aluminium = 4) with short (7 mm) carbon fibres which had been dispersed by sonication, filtered out by a wire sieve to form a preform of carbon fibres 0.15–2 mm thick. These preforms were impregnated with the geopolymer and between 20 and 50 layers laminated together using a vacuum bag, then cured at 80 °C for 24 h. Maximum toughening was obtained at 3.5–4.5 % volume fraction.

Kumar (2012) found that 0.5 wt% (of cement) of carbon nanotubes was the optimum addition level for compressive strength and splitting tensile strength. He thought that higher levels were not sufficiently dispersed even with extended sonication.

Chung (2005) investigated dispersion of short fibres in cement by measuring electrical resistivity of the composite. She found that the effectiveness of a fibre for improving functional properties was greatly affected by degree of fibre dispersion. This was particularly important when fibre volume fractions are low. Fibre dispersion is enhanced by increasing the hydrophilicity of the fibre. She investigated the effect of various admixtures (silica fume, water based polymer dispersions and solutions and silanes). Silica fume (15 wt% on cement) increased tensile strength and reduced electrical conductivity when used with 0.35 vol% carbon fibre (15 μm diameter × 5 mm long). The use of water soluble methyl cellulose (0.4 wt% on cement) had a similar effect. The use of latex and polymer dispersions also improved fibre dispersion in the region of 10–20 wt% addition on cement. The use of silica fume and methyl cellulose together had a synergistic effect on fibre dispersion.

Several other authors have investigated dispersion of discrete carbon based reinforcement in cement pastes (Manzur and Yazdani 2010; Gay and Sanchez 2010; Konsta-Gdoutos et al. 2010). They all used sonication in conjunction with a surfactant to obtain optimum dispersion of the carbon reinforcement.

Chen et al. (2011b) reviewed the current position of carbon nanotube-cement composites with respect to the fabrication, hydration, mechanical properties, electrical resistivity and transport properties of CNT-OPC composites. The authors thought that more work was required in all the fore mentioned areas to generate composites suitable for stronger and more ductile applications.

Rodriquez et al. (2010) compared the properties of two rebar reinforced concrete beams, one of which contained 1 % CNF by volume. The CNF had a diameter of 149 nm and a length of 19 μm giving an aspect ratio of 128. A polycarboxylate superplasticiser was used in both mixes and the workability was adequate to pour the two beams. The CNF increased both the strength and ductility of the beam and in addition the hardened CNF containing mix was suitable for use as a reversible strain sensor.

Work with carbon fibre to reinforce OPC began in the 1970s. Research in Japan and the USA with chopped and milled fibres based on pitch based low modulus carbon fibres has been reported on by Banthia (1994). Fibre lengths were in excess of 3 mm and had diameters of 14.5 µm and silica fume was used as a dispersing agent. Additions of carbon fibre up to 5 vol% were evaluated with increases in tensile, flexural and impact strengths reported. Improvements in water and acid resistance were obtained as the carbon fibre loading increased. These composites also showed good resistance to freeze-thaw cycling.

Banthia (1994) also reviewed applications for carbon fibre reinforced composites. One of the major uses is for thin precast products such as roofing sheets, panels, tiles, curtain walls, wave absorbers and I and L beams. CFRC curtain walls have been in use in Japan for some time. CFRC has potential for use in structures in seismic regions, for thin repairs and machinery bases. The good conductivity of these composites may be utilised in secondary anode systems in cathodic protection of reinforced concrete bridge decks.

Lyon et al. (1996, 1997, 1999) produced a woven carbon fibre-geopolymer composite with outstanding fire resistance. Tran et al. (2009) used a geopolymer based on thermal silica, kaolin and potassium waterglass to laminate a woven carbon fibre textile. The laminates contained between 37 and 40 wt% carbon fibre and were cured at a range of temperatures between 55 and 115 °C. Optimum flexural strength was obtained between 75 and 100 °C. He also claimed that the geopolymer used gave better results than one based on metakaolin. SEM showed the formation of cavities in the laminate cured above 100 °C. This was attributed to trapped water volatilising above 100 °C.

C-GRID (Chomarat 2009) is a carbon fibre-epoxy based reinforcement suitable for a range of building products from precast sections, shotcrete, and double tees. Concretes using C-GRID are characterised by light weight and high strength to weight ratio compared to steel mesh, excellent crack control and no corrosion staining.

Vera-Agullo et al. (2009) compared CNFs and MWNTs with a nanosilica (particle size 3–15 nm and a BET surface area between 20 and 1,000 m^2/g) in OPC. The nanosilica showed the best early strength development, which was attributed to the pozzolanic effect of the nanosilica. The CNF and MWCNT both increased the flexural strength at 28 days.

4.4 Inorganic Fibres

The predominant inorganic fibre used in cementitious systems is alkali resistant glass (ARG), which has a minimum zirconia content of 16 wt% zirconia (ASTM Committee C27-40 2008). Nippon Electric Glass (2007) manufacture a 19 wt% zirconia glass with outstanding alkali resistance.

The Fibreglass Reinforced Concrete Association have published a specification for the various manufacturing formats of glass reinforced concrete (Concrete

Society 2010). In addition to the conventional mortar and concrete techniques, the spray up application similar to polymer based FRP is covered.

ARG will reduce plastic shrinkage in concrete (Nippon Electric Glass 2000) and at the levels used will not affect slump resistance.

Nycon AR-DM (Nycon) is designed for dry mix applications where the fibres are exposed to highly abrasive conditions. This is achieved by a fibre coating which breaks down on water addition.

Vijai et al. (2012) added alkali resistant glass (6 mm long × 14 μm diameter) to a fly ash based geopolymer concrete at 0.1–0.03 vol% based on the concrete. In order to achieve room temperature cure 10 wt% of the fly ash was replaced by Portland cement. The 0.03 vol% fibre addition gave small increases in compressive and splitting tensile strength and a 16 % increase in flexural strength.

Fibres based on basalt are being promoted for composite reinforcement in a range of binders. Basalt is naturally occurring and less energy demanding compared to glass fibre production.

Relative to most common igneous rocks, basalt compositions are rich in MgO and CaO and low in SiO_2 and the alkali oxides, i.e., $Na_2O + K_2O$, consistent with the total alkali silica (TAS) classification. The TAS classification is used to classify volcanic rocks (Australian Museum 2007).

Basalt generally has a composition of 45–55 wt% SiO_2, 2–6 wt% total alkalis, 0.5–2.0 wt% TiO_2, 5–14 wt% FeO and 14 wt% or more Al_2O_3. Contents of CaO are commonly near 10 wt%, those of MgO commonly in the range 5–12 wt%.

De Fazio (2011) suggested that only basalt with silica contents greater than 46 % were suitable for fibre formation. Also the presence of iron improved acid resistance. The presence of iron oxides in these basalt fibres showed improved chemical and heat resistance compared to standard, E-glass fibres. This work rated basalt fibres behind ARG (with added zirconia) with regards to alkali resistance.

Lipatov et al. (2012) investigated the effects of adding zirconia to basalt fibres at the manufacturing stage. They concluded from XRF results that the maximum solubility in basalt was 7.1 wt%. XRD showed that the zirconia was present in the tetragonal and monoclinic forms. Optimum alkali resistance (2 M NaOH for 3 h at 98 °C) was obtained at 4 wt% added zirconia (3.1 wt% in the fibre). At 4 wt% added zirconia, tensile strength of the fibres decreased by more than 50 %.

Jung and Subramanian (1994) coated basalt fibres with zirconium-n-propoxide stabilised by chelation with ethyl acetoacetate or without stabilisation. Alkali attack on the uncoated fibres results in dissolution of SiO_2, Al_2O_3, and CaO with the formation of insoluble hydroxides of iron, magnesium and titanium from the constituents of the basalt. These reactions are suppressed by the unstabilised zirconia coating on the basalt fibre. The tensile strength of uncoated fibres is drastically reduced by alkali attack, whilst the coated fibres maintained their strength after 90 days immersion.

Scheffler et al. (2009) investigated the alkali resistance of basalt and alkali resistant glass fibres in 5 % NaOH (pH > 14) and filtered cement solutions (pH = 12.8–12.9). The attack on the fibres by NaOH leads to a strong dissolution of the outer layers of both fibres with the formation of an extensive shell like layer and a reduction in fibre diameter. In cement solution the attack is more localised shown by the formation of holes of varying size and no reduction in fibre diameter. An insoluble calcium rich surface layer appears to inhibit further attack on the fibre surface. When sizes based on epoxy film formers and silanes are applied to the fibres less shell like layer was formed suggesting improved NaOH resistance.

Dias and Thaumaturgo (2005) investigated the addition of basalt fibres (45 mm × 9 µm) to metakaolin based geopolymer and OPC at 0.5 and 1 % volume fraction. The basalt fibres were more efficient at reinforcing the geopolymer system. This could probably be related to the bond between fibre and matrix.

Kriven et al. (2008) used a metakaolin geopolymer with a nominal SiO_2:$Al_2O_3 = 4$:1 as a binder for woven basalt fabric and chopped basalt fibre (4 mm long × 10 µm diameter). Beyond 1 % chopped fibre addition the paste became un-pourable. Using 6 layers of woven basalt fibre, three point bend strength and work of fracture increased from 2.75 to 10.25 MPa and from 0.05 to 21.82 kJ m^{-2}, respectively. Also in this work stainless steel powder was added to the geopolymer with increases in compressive strength and flexural strength up to 15 wt% addition (note: this is about 1 % by volume of stainless steel). It was suggested that the stainless steel particles were blunting propagating cracks.

Rill et al. (2010) added chopped basalt fibres (6 mm long × 13 µm diameter) at up to 10 wt% to a metakaolin based geopolymer (activated with potassium and SiO_2:$Al_2O_3 = 4$:1). A 10 wt% addition was the limit due to workability issues. Adding the fibres in an IKA shear mixer or a Thinky centrifugal mixer caused the fibres to break down to about 100 µm long. The fibres were therefore blended in by hand. The flexural strength increased from 1.75 to 19.5 MPa (10 wt% fibres). The basalt fibres had been supplied with a silane size. In a separate evaluation the size was removed by calcination at 600 °C for 1 h. When this material was added to the geopolymer at 1 wt% the flexural strength was 2.4 MPa compared to 3.6 MPa for sized fibres. The presence of the sizing helped the workability by reducing clumping during mixing. Samples containing 7 wt% fibres were heated to 500 and 1,000 °C. The 500 °C sample exhibited micro-cracking but showed a flexural strength of 4.7 MPa (16.5 MPa for unfired). The 1,000 °C sample showed high shrinkage and was too weak to test. These observations were attributed to dehydration and crystallisation of the geopolymer at the higher temperature. No SEM was undertaken to investigate changes in the basalt fibres at elevated temperatures.

Fibres Unlimited (2007) compared basalt fibres with polypropylene and polyacrylonitrile fibres with respect to workability and final properties in OPC.

Fellicetti et al. (2001) investigated two basalt fibre grades for down hole cementing slurries. One grade proved superior in alkali resistance and was used for the balance of the work. In down hole cementing flowability is critical and evaluation with a mini cone was carried out with 10 µm diameter fibres. A 1–2 % volume addition of 1 mm long fibre gave optimum flow results. Longer fibres (up to 5.5 mm long) gave pronounced improvements in composite toughness.

Many new supply forms and applications based on basalt are currently available (Kamenny Vek 2012; ReoforceTech 2012a, b). The main areas of usage are reinforcement in composites, high temperature insulation and fire protection.

4.4 Inorganic Fibres

The 3M company manufacture a range of alumina fibres under the brand name Nexel (3M 2003). They have increasing crystalline Al_2O_3 contents with corresponding lower amorphous silica contents. The top end grades (Nexel 610 and 720) are completely crystalline with melting points of 2,000 and 1,800 °C, respectively. Individual fibre diameters are in the 10–12 μm range. Nexel 610 and 720 are used for load bearing composite applications over a wide temperature range.

Foerster et al. (1994) added alumina particles, fibres and woven fabric to geopolymer (Geopolymite). Short fibres (150 μm long × 3 μm diameter) gave a rapid increase in viscosity above 5 vol%, with increases in toughness and compressive strength. The addition of particles (fillers) gave much slower increases in viscosity and strength up to 25 vol%. A 2 ply woven alumina fabric laminate (35 vol% alumina) only provided 50 % of the stiffness of a carbon fibre rovings laminate (75 vol% carbon). However, this was double the stiffness obtained with 6 vol% of short alumina fibre in the binder. The authors also reported that Nexel 312 alumina weaves (a product of 3 M) were badly degraded with loss of strength.

Defazio et al. (2006) evaluated three alumina ceramic supply forms, alumina tissue/paper, milled alumina fibres, and short, 3 mm, fibres (chopped). Three manufacturing techniques were utilised, hand layup combined with vacuum bagging, pour/vibrate, and vacuum bagging followed by curing under pressure in a heated press at 80 °C. The geopolymer was varied from $SiO_2:Al_2O_3$ ratios of 1:1 to 5:1, but no other compositional information was given. The primary objective was to develop a composite with an operating temperature of 1,350 °C and a flexural strength of 75 MPa. Maximum flexural strength (97 MPa) was obtained using the heated press and 11.3 % loading of short ceramic fibres.

Thang et al. (2010) added alumina nanofibres to a geopolymer matrix (Q17 from the Geopolymere Institute). This blend was now used as a laminating binder system for carbon fibre, basalt fibre and glass fibre unidirectional rovings. Additions of nanofibres in 0.5 wt% steps from 0.5 to 2 wt% were made. Flexural strength values peaked at 1 wt% with each of the unidirectional fibres. The authors believe that a dispersing and/or bonding agent for the nanofibres could be of great interest.

Nazari et al. (2010) added alumina nanoparticles (average diameter 15 nm) to OPC concrete in 0.5 wt% steps up to 2 wt%. Initial setting time was reduced by 50 % and final setting time by 33 %. Only small increases in splitting tensile strength and flexural strength were seen and the authors thought that "needle like particles" would be more beneficial. However, these nano particles were having a beneficial influence on cement hydration.

Bernal et al. (2012a) investigated the effects of milled recycled refractory brick and an alumina/silica/zirconia fibre at elevated temperatures in a metakaolin geopolymer. Unfilled geopolymer showed high shrinkage and cracks due to severe dehydration above 200 °C. Filled samples were pre-dried at 65 °C prior to further heating. The recycled aluminosilicate brick was used to even out the differences in thermal expansion and reduce formation of cracks. The addition of the fibres led to workability issues which the authors believe lead to high variability of the strength test results. However, the presence of fibres controlled the formation of visible cracks.

Fig. 4.8 Wollastonite fibres (Nyad MG)

Silicon carbide based fibres and woven fabrics are available commercially (COI Ceramics 2006). Davidovits (1991) reported on the use of woven silicon carbide fabrics in geopolymer with superior high temperature thermal strength compared to E-glass and carbon fibre composites.

Silva et al. (1999) used naturally occurring wollastonite fibres (aspect ratio = 10–20:1) to reinforce a metakaolin based a geopolymer. Toughness of the system increased as fibre volume increased to 5 vol%. The wollastonite is compatible with the pH levels in geopolymer synthesis and a dense interfacial transition zone, matrix-fibre, was developed.

Figure 4.8 shows an SEM image of commercially available wollastonite, Nyad MG ex Nyco Minerals. The acicular nature and broad size distribution is clearly visible.

Chapter 5
Thermal Properties of Geopolymers

Abstract The physics and testing methodology of thermal properties are introduced prior to a review of OPC and geopolymer thermal properties. The amorphous inorganic structure of the geopolymers lends itself to good thermal resistance which leads to potential applications such as thermal insulation. Thermal expansion can generate destructive internal stresses when structural parts are heated and restrained from moving. The thermal expansion measuring techniques commonly utilised are dilatometry, interferometry and thermomechanical analysis (TMA). Thermal expansion measurements of metakaolin and fly ash based geopolymers show several distinct regions as the temperature increases. The extent of these regions varies from system to system and the changes are attributed to dehydration, dehydroxylation, densification and crystallisation. Fillers and aggregates can be added to geopolymers to reduce the thermal expansion of the composite and extend the usable temperature range. Thermal conductivity determination is required to assess geopolymers' suitability for potential applications in thermal barriers and construction structural members. The two approaches to measuring thermal conductivity: steady state and transient (non-steady state) techniques are compared. The microstructure of geopolymer profoundly influences thermal conductivity; particularly porosity which if increased leads to a reduction in thermal conductivity. The addition of aggregate influences the thermal conductivity of OPC and geopolymer concrete and at the same time decreases thermal durability due to mismatch of thermal conductivity between aggregate and matrix. Some geopolymers with low initial strength have been shown to gain strength after exposure to high temperatures. It has been hypothesised that unreacted precursor levels can convert to geopolymer at high temperature and increase the strength. Once again the importance of the geopolymer microstructure is highlighted.

Keywords Thermal expansion · Thermal conductivity · Microstructure of geopolymer · Thermal incompatibility · Fillers and aggregates

Geopolymers based on metakaolin and fly ash are considered in this section. The amorphous inorganic structure of the geopolymers lends itself to good thermal resistance which leads to potential applications as thermal insulation in industrial, military and domestic sectors. Fire resistance is a large subset of thermal properties and will be considered in a dedicated section.

Thermal properties of interest include:

- Thermal expansion and shrinkage
- Thermal conductivity
- Mechanical strength retention
- Creep
- Explosive spalling
- Microstructure including phase stability and resultant changes, pore size and distribution, weight changes including dehydration and rehydration and related thermochemical properties.

Thermal analysis techniques such as, thermogravimetric analysis (TGA), differential scanning calorimetry (DSC), and differential thermal analysis (DTA), can be used to gather information on phase stability and thermodynamic properties of materials.

The composition of the geopolymer will have an influence on performance at elevated temperatures. The ratio of silicon to aluminium (Si:Al) plays a large part in the development of geopolymer properties and the influence of heat on these properties. This ratio should refer to the material contained in the polymer (binder ratio) and not the precursors (overall ratio) as complete conversion to geopolymers rarely occurs. Some authors use the $SiO_2:Al_2O_3$ ratio which will give a higher numerical value than the Si:Al ratio for the same material. These nomenclature issues are in need of resolution to bring more clarity to geopolymer research.

5.1 Measurement of Thermal Transport Properties

The analysis of heat transfer through structures is of great importance in civil engineering problems such as heat flow into buildings in energy efficient designs, thermal loadings of buildings due to daily temperature variations, design of buildings for thermal comfort, radiation shield design in nuclear power stations, analysis of bridge decks and other surfaces exposed to solar thermal loading (Khan 2002).

The word heat should only be used to describe energy transfer from one place to another. Heat flow is an energy transfer that takes place as a consequence of temperature differences only (Serway 1992).

The heat capacity, c, of a specific specimen of a substance is defined as the amount of heat energy needed to raise the temperature of that specimen by 1 °C (Serway 1992).

If Q units of heat cause a temperature change of ΔT then:

$$Q = c \cdot \Delta T \tag{5.1}$$

5.1 Measurement of Thermal Transport Properties

The heat capacity of a specimen is proportional to its mass. It is convenient to define the heat capacity per unit mass of substance, as the specific heat, C:

$$C = \frac{c}{m} \tag{5.2}$$

Latent heat is the energy required to change the phase of a substance e.g. solid to liquid or liquid to gas or changes to crystalline structures. The temperature does not rise during a phase change.

Specific heat and latent heat may be determined by differential scanning calorimetry (DSC). DSC measures the difference between heat flows from the reference and sample sides of a sensor as a function of time or temperature. Differences in heat flow occur when a sample absorbs or releases heat due to thermal events such as melting, crystallisation, and chemical reactions.

Heat transfer processes include conduction, convection and radiation. The conduction of heat only occurs if there is a temperature difference between two parts of the conducting medium. If we have a slab of cross sectional area, A and thickness, Δx, with opposite faces at temperatures T_1 and T_2, the heat flow, H, is found to be proportional to the temperature difference, and the cross sectional area and inversely proportional to the thickness. We can write:

$$H = -kA\frac{dT}{dx} \tag{5.3}$$

k is the thermal conductivity of the material with units of W m^{-1} K^{-1}.

Table 5.1 lists thermal conductivity values for metals, non-metals and gases. Metals typically have high thermal conductivities due to the high number of electrons that are relatively free to move and transport energy over large distances. Non-metals have covalent or ionic bonds which conduct heat predominantly by phonon vibrations, whilst gases show no order with high separation distances between molecules at ambient temperatures giving rise to few collisions able to transfer energy.

There are two approaches to measuring thermal conductivity: steady state and transient (non-steady state) techniques are outlined below.

Table 5.1 Thermal conductivity, k, values adapted from Serway (1992)

Metals at 25 °C	K (W m^{-1} K^{-1})	Gases at 20 °C	K (W m^{-1} K^{-1})	Non-metals	Approximate values of k (W m^{-1} K^{-1})
Aluminium	238	Air	0.0234	Asbestos	0.08
Copper	397	Helium	0.138	Concrete	0.8
Gold	314	Hydrogen	0.172	Glass	0.8
Iron	79.5	Nitrogen	0.0234	Ice	2
Lead	34.7	Oxygen	0.0238	Rubber	0.2
Silver	427			Water	0.6
				Wood	0.08

In steady state techniques a measurement is performed when the temperature of the measured specimen does not change with time. A solid sample of known dimensions is placed between two temperature monitored plates. One plate is heated whilst the other plate is cooled and their temperatures monitored until they are constant. The thermal conductivity is calculated from the steady state temperatures, specimen dimensions and heat input.

Sample preparation can be extensive and testing time will be in the range of hours. Analysis of outputs is relatively straight forward. The guarded hot plate is a commonly used steady state test procedure as outlined in ASTM C518-10 and ASTM C177-10 (ASTM Committee C16.30 2010a, b).

In transient techniques (hot wire method) as per ASTM D5930-09 (ASTM Committee D20.30 2009) a measurement is made during the heating process of the specimen and measurements can be recorded relatively quickly. The measurements are plotted as a function of time. Whilst the testing period is short, mathematical analysis of the data is generally more complex.

The hot wire method involves inserting a wire into the sample either during the moulding process or by a post cure machining technique. A current passed through the wire generates heat which flows out radially from the wire and the temperature (T) of a fixed point close to the wire is recorded by a thermocouple inserted into the sample (Glatzmaier and Ramirez 1988; Healy et al. 1976).

$$\Delta T(t,r) = \frac{Q}{4\pi k}\left\{-\gamma - \ln\left(\frac{r^2}{4at}\right) + \ln(t)\right\} \quad (5.4)$$

where

ΔT = Temperature at time t – initial temperature;
t = time of heating;
r = radial distance to line source;
Q = power per unit length;
k = thermal conductivity of sample;
γ = Euler constant (0.577);
α = thermal diffusivity.

By plotting the temperature of this fixed point verses the natural logarithm of time, thermal conductivity can be measured from the slope given knowledge of the input power, Q (refer to Eq. 5.4).

A variation of the hot wire method is the transient plane source method that employs two sample halves, in between which the sensor is sandwiched. The sensor consists of a double spiral of electrically conducting nickel metal which is encapsulated in thin polyimide film, which provides electrical insulation and mechanical protection. During testing a constant electrical signal is passed through the conducting spiral, increasing the sensor temperature. The heat dissipates into the specimen on either side of the sensor at a rate dependent on the thermal transport properties of the specimen. Typically temperature increases are less than 2 °C. Recording of time and temperature in the sensor allows the thermal properties to be calculated. The rate of increase of the sensor voltage is used to determine thermal conductivity, thermal diffusivity and specific heat capacity of the specimen.

The modified transient plane source method uses a heating element that is supported on the back side by an insulating material thus allowing one directional heat flow. The equipment uses a one sided, interfacial heat reflectance sensor that applies a momentary constant heat source to the sample. This modification provides one sided facial measurements of liquids, powders, pastes and solids. Little if any sample preparation is required and testing time is seconds (C-Therm 2010).

The laser flash method is based on the measurement of the temperature rise at the rear face of a thin disc specimen. The temperature rise is produced by a short energy pulse on the front face. An infra-red detector is used to measure temperature. Sample preparation is extensive and operator dependent but measuring times are short. This method can be used for evaluations up to 2,000 °C (Kamseu et al. 2012a).

Samples with low emissivity or absorptivity are coated with a graphite film to increase energy absorbed on the laser side and increase temperature signal on the back side. Infrared radiation transparent materials must be coated with a metal film (typically 0.1 μm gold) on both sides to prevent penetration of the laser beam at the front and to prevent infrared detectors viewing into the sample at the back.

$$k = \rho \cdot Cp \cdot \alpha \qquad (5.5)$$

where

k = Thermal conductivity,
ρ = density,
C_p = heat capacity (from DSC),
α = thermal diffusivity.

Substances with high thermal diffusivity rapidly adjust their temperature to that of their surroundings because they conduct heat quickly in comparison to their volumetric heat capacity. Thermal diffusivity can be measured by the laser flash method.

Thermal conductivity is the property that determines how much heat will flow in a material, while thermal diffusivity determines how rapidly heat will flow within the material.

A sample's thermal resistance (R value) is equal to its thickness, x, divided by its thermal conductivity, k. Thermal resistance measures the ability of materials to resist heat transfer (Kamseu et al. 2012a).

5.2 Thermal Expansion

The coefficient of linear expansion (COTE or α) is a material property that is indicative of the extent by which a material expands or contracts on heating. The value varies over the wide range of available materials and over small temperature ranges is proportional to temperature change. Thermal expansion can generate destructive internal stresses when structural parts are heated and restrained from moving.

$$\alpha = (\Delta L/L_i)/\Delta T \tag{5.6}$$

where

L_i = initial length;
ΔL = change in length;
ΔL = change in temperature.

Equation 5.6 (Serway 1992) assumes α is constant over the temperature range. However, this is only true over limited temperature ranges. For a solid the coefficient of volume expansion, β, is three times α. This assumes that the material is isotropic.

To determine COTE the displacement and temperature must be measured on a specimen that is exposed to a thermal cycle. The main techniques are dilatometry (ASTM Committee E37 2011), interferometry (ASTM Committee E37 2010) and thermomechanical analysis (TMA) (ASTM Committee E37 2012).

In the case of dilatometry a uniform section specimen is heated in a furnace and displacement of the ends of the specimen are transmitted to a sensor by means of push rods. Push rods can be made from vitreous silica, high purity alumina (service temperature up to 1,600 °C) and isotropic graphite (up to 2,500 °C). Compensation for the expansion of the push rod material during the test heating cycle has to be allowed for (ASTM Committee E37 2011). The setup of dilatometer experiments is kept as force free as possible. Dilatometer measurements are generally applicable to materials with linear COTE of more than 0.5 μm m^{-1} K^{-1}.

With optical interference methods a sample of known geometry can be given polished reflective ends or placed between flat reflective surfaces (mirrors). The mirrors are flat uniform thickness pieces of silica or sapphire with surfaces partially coated with gold or other reflective metal. Light either from a laser or monochromatic source illuminates each (end) surface simultaneously to produce a fringe pattern. As the sample is heated (or cooled) expansion (or contraction) causes a change in the fringe pattern due to the optical path length differences between the end reflecting surfaces. This change is detected and converted into a length change from which COTE may be determined. Interferometry is normally used in the range of −150–700 °C. Precision at ± 40 nm m^{-1} K^{-1} is significantly better then dilatometry or thermomechanical analysis techniques. This technique is suitable for materials having low (<5 μm m^{-1} K^{-1}) or negative COTE (ASTM Committee E37 2010).

Unlike traditional methods such as the dilatometer, in thermomechanical analysis (TMA) measurements a constant small load (0.1–5 g) is applied to the specimen via a vertically adjustable quartz glass probe. This probe is integrated into an inductive position sensor. As the sample expands (or contracts) under heating it moves the probe and the recorded signal manipulated to give COTE. A range of probes are available for TMA to allow measurements of COTE, penetration for T_g determination, a tension jig for films and fibre evaluations. Testing of non-uniform samples and powders is possible by immersing the sample in a known volume of liquid, typically silicone oil, in a suitable retaining vessel which is attached to a suitable probe. Figure 5.1 shows several probes used in TMA (Anasys Thermal Methods Consultancy 2012).

5.2 Thermal Expansion

Fig. 5.1 Different probes used in TMA (Anasys Thermal Methods Consultancy 2012)

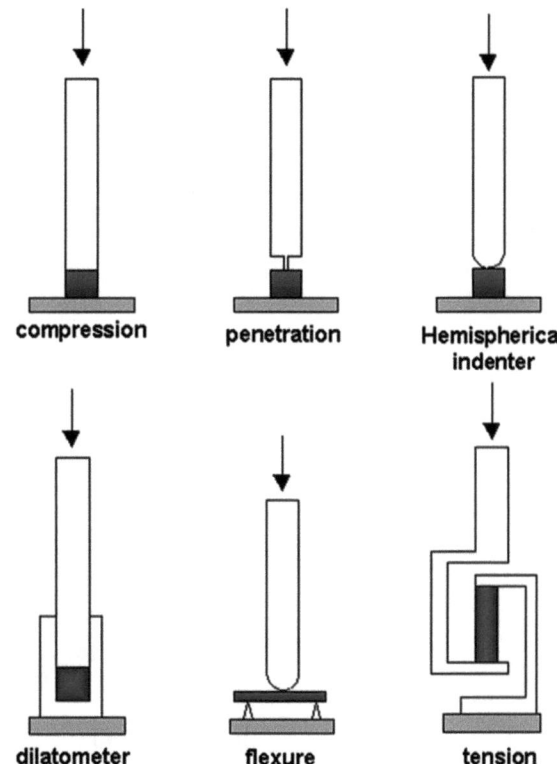

5.2.1 Thermal Expansion of Geopolymers

Barbosa and MacKenzie (2003a, b) made geopolymers from metakaolin and either sodium hydroxide or potassium hydroxide and measured the thermal resistance of the resulting materials. They also investigated the addition of 10 vol% of various inorganic fillers on strength and thermal properties and the influence of water content of the geopolymer formulation. Water contents ($H_2O:Na_2O$) of 10, 17.5 and 25 were evaluated, but only the lower water content cured sufficiently for dilatometer and strength testing. The results are shown in Fig. 5.2.

The TG curve (A) showed that the well cured samples retain 15 wt% water of which 12 wt% is lost below 230 °C, the balance is either chemically bound or unable to migrate out of the sample readily. However, it continues to evolve gradually up to 500 °C. The DSC curve shows that the energy associated with the endothermic water loss is -0.24 kJ g^{-1}.

The dilatometer curve (B) shows that the water loss is associated with a small shrinkage value, and the material is essentially stable between 250 and 800 °C. Above 800 °C further shrinkage takes place due to densification and/or volume changes as a result of crystallisation and subsequent melting. The shrinkage stops at 880 °C and becomes dimensionally stable up to 1,000 °C, the limit of the experiment. The sample retains this stability on cooling and during further thermal cycling.

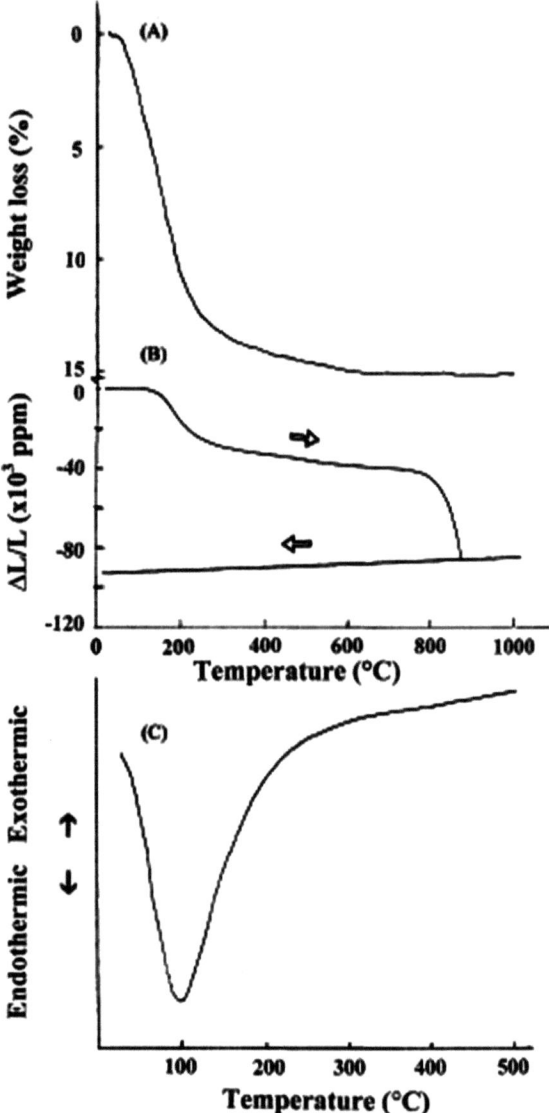

Fig. 5.2 TG (*curve A*) at 20 °C min^{-1}, dilatometer (*curve B*) at 1 °C min^{-1} and DSC (*curve C*) at 10 °C min^{-1} (Barbosa and MacKenzie 2003b)

The addition of 10 vol% of inorganic filler to the geopolymer binder resulted in a drop in compressive strength with the samples containing alumina and β-sialon showing the most marked decrease in strength. This was attributed to both of these fillers being unreactive at ambient temperatures and unable to bond with the binder. They dilute the geopolymer without contributing to the strength of the composite. Dilatometer testing of these filled composites showed a flat portion in the curves up to 800 °C indicating an upper service temperature similar to pure geopolymer.

The authors also commented that microcracking after demoulding could be reduced by the addition of up to 5 wt% glycerol, which plasticised the mix.

5.2 Thermal Expansion

Rahier et al. (1997) used thermomechanical analysis to study metakaolin geopolymers produced from sodium or potassium silicates of varying moduli. All samples showed shrinkage due to dehydration up to around 230 °C. A second shrinkage event was dependent on the silicate modulus and alkali metal cation type. In the case of sodium silicate an increase in modulus from 1.4 to 1.9 resulted in the onset temperature increasing from about 650 to about 800 °C whilst changing the cation from sodium to potassium, at a modulus of 1.4 resulted in an onset temperature of about 900 °C.

Duxson et al. (2006a) used metakaolin geopolymers and investigated the influence of initial curing conditions on thermal shrinkage. He compared samples cured at 80 °C and those matured at 20 °C for 24 h prior to curing at 40 °C. The specimen cured at the higher temperature began to shrink at a lower temperature but is ultimately more stable and densifies at a higher temperature. The sample matured prior to curing at 40 °C showed the same properties as that only cured at 20 °C. He believed that this indicated that thermal shrinkage outcomes were determined by curing temperature and not pre-maturation.

Duxson et al. (Duxson et al. 2006b; Duxon et al. 2007a) carried out detailed work on metakaolin based geopolymers for a range of silicon:aluminium ratios using sodium and potassium silicates and their blends, measuring the thermal shrinkage of these materials and introducing the concept of four temperature regions to characterise the changes taking place:

Region I involves the loss of freely evaporable water with only minimal shrinkage.
Region II shows the start of initial shrinkage and continued significant weight loss.
Region III is characterised by gradual weight loss and thermal shrinkage.
Region IV involves densification and sintering and is characterised by rapid shrinkage and small weight loss.

The mechanism of shrinkage in region II is thought to be capillary shrinkage from evaporation of free water with the extent of shrinkage increasing with water content. The rate of shrinkage in region III was similar in all specimens regardless of silicon:aluminium ratio. However, the temperature span of region III was related to the silicon:aluminium ratio by dehydroxylation. The high temperature required for condensation of aluminol groups is thought to increase the temperature required for complete dehydroxylation in specimens with low silicon:aluminium ratios. The onset temperature of region IV decreased with increasing silicon:aluminium ratio due to incomplete incorporation of aluminium from the metakaolin providing free sodium cations.

Rickard et al. (2010) expanded the above as shown in Fig. 5.3 and Table 5.2 to include characteristics specific to fly ash based geopolymers. The temperature range for each region is variable and dependent on sample composition and heating rates.

Geopolymers expand on heating (region I), however they contain varying amounts of water that exist either adsorbed in the pores or chemically bound in the structure. On heating the water dehydrates at various stages depending on the energy requirements. Dehydration leads to a shrinkage event and the overall thermal expansion is a convolution of the solids expansion and the shrinkage of the pores (region II).

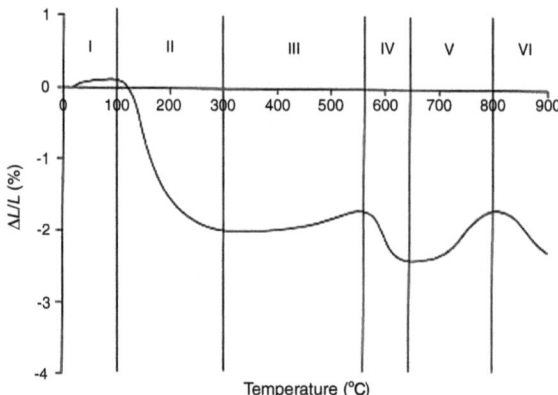

Fig. 5.3 Thermal expansion of a fly ash geopolymer (Si:Al = 2.3, w/c = 0.2) (Rickard et al. 2010)

Table 5.2 Thermal shrinkage/expansion characteristics of geopolymers (Rickard et al. 2010)

Region	Temperature range (°C)	Description	Effect	Factors
I	0–150	Resistive dehydration	Slight expansion	Young's modulus of sample. Heat rate
II	100–300	Dehydration of free water	Significant shrinkage	Water content Heat rate
III	250–600	Dehydroxylation	Minimal shrinkage	Abundance of hydroxyl groups and chemically bound water
IV	550–900	Densification by vitreous sintering	Significant shrinkage	Residual water content Si:Al ratio
V	Above densification temperature	Crystallisation in the geopolymer paste. Expansion due to cracking	Moderate expansion	Compositional ratio Concentration and type of impurities
VI	Above densification temperature	Further densification	Large shrinkage	Compositional ratio

The slight thermal shrinkage occurring in region III between 300 and 600 °C is due to physical contraction as the hydroxyl groups are released creating shorter T–O–T links. The small amount of shrinkage can be masked by the expansion of secondary phases particularly the high concentration of impurities found in fly ash.

$$T-OH + HO-T\equiv \;\rightarrow\; \equiv T-O-T + H_2O \tag{5.7}$$

where T is an aluminium or silicon atom.

5.2 Thermal Expansion

The second major shrinkage event occurs between 550 and 900 °C (region IV) due to densification of the geopolymer as the paste sinters and viscous flow fills any voids present in the material. Duxson (2007a) found that the presence of residual water in the material after dehydration reduces the activation energy for viscous flow.

Beyond the densification region (region V) no consistent trends are observable. The differences in thermal expansion in this region are believed to be due to compositional differences and the presence of impurities in precursor material. Rickard et al. (2010) and Rahier et al. (1997) reported expansive behaviour, whilst Duxson et al. (2007a) and Dombrowski et al. (2007) saw thermal shrinkage events. Barbosa and MacKenzie (2003a) stated that samples were thermally stable. The formation of cracks and changes to the nature of the pore structure also influence thermal expansion in region V.

In region VI the thermal expansion is characterised by large and rapid shrinkage events. The causes of this shrinkage may be attributed to continued densification, destruction of crystalline phases formed in region V, collapse of the pore system or sample melting.

Dombrowski et al. (2007) evaluated the effect of added calcium to fly ash geopolymers and found there was an optimum level at 8 wt% calcium hydroxide for both mechanical strength and heat resistance. He set 2 % for the acceptable maximum for thermal linear shrinkage. The 8 wt% calcium content mix reached this value at 1,050 °C compared to 810 °C and 960 °C for 0 and 20 % respectively. He also carried out a creep test at 960 °C for 25 h under compression where the 8 wt% mix did not shrink below 2 %. The 20 wt% mix dropped just below 2 % whilst the 0 wt% mix dropped to 6 %. He claimed that the 8 wt% calcium lead to the formation of more crystalline compounds whilst at 20 wt% the calcium depressed the melting point. The 8 wt% calcium sample showed the highest concentration of nepheline at 800 °C and feldspar at 1,000 °C, which contributed to highest strength and lowest shrinkage.

Subaer (Subaer and van Riessen 2007) investigated the addition of aggregates to metakaolin based geopolymers and their influence on thermal properties. The water loss of about 15 wt% from ambient to 250 °C gave rise to approximately 2 % shrinkage. Between 250 and 800 °C the sample was thermally stable, indicating the maximum service temperature. Beyond 800 °C further shrinkage occurred due to densification. The foregoing data is for a geopolymer with silicon:aluminium = 1.5 and sodium:aluminium = 0.6. Adding 20 % wt% aggregate to the geopolymer reduced the shrinkage to around 1 % in the range of 23–500 °C. A further reduction in shrinkage was obtained by increasing the aggregate to 40 wt%.

Geopolymers containing quartz aggregate show an abrupt expansion between 500 and 540 °C followed by further shrinkage as the temperature increases. There is a low-high quartz transition at 574 °C (The quartz page 2012). The discrepancy arises from the geopolymer being hotter in the interior due to its exothermic nature as indicated by DSC. This transition is the reason fire resistant OPC is never produced from quartz aggregate (Neville 1995). Granite filled geopolymer is stable up to 800 °C.

Lin et al. (2009b) added up to 30 wt% of 0.75 μm α-alumina to a metakaolin geopolymer which had been synthesised by reaction with potassium silicate.

Using TG-DTA they demonstrated that α-alumina had little effect on the thermal evolution of geopolymers at high temperatures. The endotherm up to 300 °C is attributed to water loss. The endothermic peaks at 958 °C (no added alumina) and 985 °C (~20 wt% alumina) were attributed to crystallisation of the geopolymer.

XRD patterns of these two systems showed a characteristic amorphous hump at ~28° (2θ using copper Kα radiation) in samples exposed up to 600 °C. The sharp reflections from leucite ($KAlSi_2O_6$) are seen in the alumina free geopolymer at 800 °C and remained until 1,400 °C. However, in the case of the 20 wt% alumina containing geopolymer leucite only appeared at 1,000 °C. From this Lin concluded that the presence of alumina increases the crystallisation onset temperature and also the rate of crystallisation.

Lin also measured the thermal volume shrinkage of specimens after exposure to elevated temperature and saw significant reductions in shrinkage values above 20 wt% alumina additions. This was particularly marked above 800 °C. Density and porosity showed marked changes above 600 °C. The higher alumina contents showed increased porosity. Flexural strength was independent of alumina content, but increased markedly in all samples above 600 °C. The fracture surface of geopolymer samples heat treated at 1,000 °C became rougher with increasing numbers of pores. Lin believed that this showed alumina has a negative influence on densification of geopolymers at high temperatures, but that it is beneficial for reducing thermal shrinkage.

Kamesu et al. (2010) made potassium activated metakaolin geopolymers which were extended with fine particle size quartz or α-alumina to evaluate the thermal properties of the blends. The total shrinkage in the range 25–900°C was less than 3 %. The maximum shrinkage of the unfilled geopolymer was 17 % recorded at 1,000 °C which was reduced to 12 % by addition of alumina. The temperature where maximum densification occurred in the unfilled geopolymer was 1,000 °C which shifted to 1,150 and 1,200 °C for 75 wt% silica and alumina filled blends, respectively. Addition of water to obtain processable blends was between 34 and 44 wt%. No physical property data was reported for the investigated blends.

Rickard et al. (2010) produced geopolymers from fly ash containing 20 wt% α-quartz and 15 wt% iron oxide impurities. Expansions on either side of the quartz high-low transition at 574 °C were minor in nature due to only 15 wt% of the geopolymer being quartz. The quartz did not react with the geopolymer as the temperature increased and therefore did not influence post heating morphology or phase composition.

On the other hand the iron oxide content directly influenced the thermal expansion by altering the phase composition and morphology after heating. After heating in air at 900 °C a portion of the amorphous iron oxide from the fly ash converted to hematite. This was confirmed by weight increases in TGA and the appearance of hematite in XRD patterns. A thermal expansion event observed in region 5 (Fig. 5.3) was attributed to the formation and subsequent growth of cracks in the specimens. This thermal expansion event had not been observed in either fly ash or metakaolin geopolymers previously and was attributed to the high iron oxide content.

5.2 Thermal Expansion

Bakharev (2006) activated two Australian fly ashes with sodium and potassium based activating solutions. One of the fly ashes contained 13 wt% iron oxides. The author thought that the large shrinkage and loss of compressive strength on firing to 1,000 °C could be attributed to the presence of these iron oxides. Shrinkage increased with increased water to binder ratio, increased alkali content and increasing the firing temperature from 800 to 1,000 °C. Melting occurred at a lower temperature when the alkali content increased. The second fly ash, with high silica content, showed thermal expansion which resulted in the formation of foam at 1,200 °C. Sodium activated fly ash gave geopolymers which exhibited a rapid deterioration in strength at 800 °C. This was connected to a rapid increase in pore size. However, geopolymers produced from potassium activated fly ash remained predominantly amorphous up to 1,200.

Pan et al. (2009) exposed fly ash based geopolymer mortars to 800 °C and found that strength after exposure sometimes decreased and at other times increased. Specimens produced from two fly ashes showed strengths in the range of 5–60 MPa. The strength loss decreases were reduced with increasing ductility, with strength gains evident at the highest levels of ductility. This was attributed to mortars with high ductility having a high capacity to accommodate thermal incompatibility. Two opposing mechanisms are thought to occur in mortars:

- Further geopolymerisation and/or sintering at high temperatures leading to strength gains.
- Damage to the mortar because of thermal incompatibility arising from non-uniform temperature distribution.

The strength gain or loss occurs depending on the dominant process.

Thermal incompatibility occurs due to heat flow in solids taking time to reach a steady state. The thermal incompatibility in non-homogeneous, multi-phase materials e.g. concrete or mortar occurs because of different expansive responses to the applied heat.

Ductility of specimens was calculated from the compressive stress-strain curves. The mortars with lower initial strength exhibited greater ductility as evidenced by a rounder shaped curve. Mortars with the higher initial strength show a narrower curve with a different response in the descending part and exhibited lower ductility.

Ductility was defined by two methods as shown in Fig. 5.4. One method is to divide strain ε_1 by strain ε_2. Strain ε_1 represents a strain at the limit of elastic behaviour. Strain ε_2 corresponds to 0.85 of the peak stress in the descending part of the curve. Another method to define the ductility index is by dividing the total energy at failure by the elastic energy stored at peak load (Pan et al. 2009).

Pan and Sanjayan (2012) were able to measure stress-strain properties in-situ at elevated temperatures. Geopolymers based on class F fly ash were prepared by activation with sodium, potassium and sodium/potassium blend solutions at various addition levels. The softening point of the sodium based geopolymer remained the same (610 ± 20 °C) regardless of silicate concentration, fly ash composition and test load. Potassium based geopolymers softened at 800 °C and

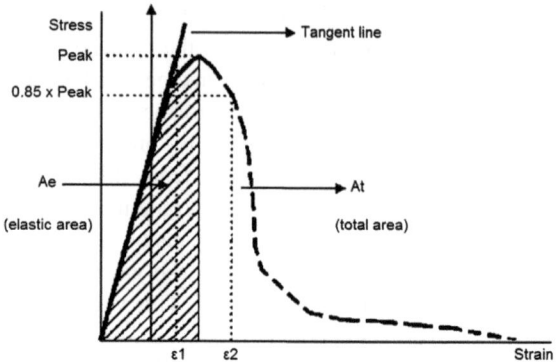

Fig. 5.4 Derivation of ductility index (Pan et al. 2009)

the sodium potassium blend at 570 °C. Increases in strength for sodium based geopolymers were seen at 530 °C and at 730 °C for potassium based geopolymers. This increase in strength was seen in conjunction with heat release indicating that an exothermic event was occurring. The author suggested that further geopolymerisation reactions were occurring at between 300 and 500 °C accounting for the increase in strength.

Two types of test were carried out:

1. Study of the deformation of heated geopolymers under constant load. Tests were carried out at stress levels of 0.05 σ_r and 0.55 σ_r (σ_r is the reference strength at ambient temperature). Deformation was measured until 83 % of the initial height was reached.
2. Compressive strength at elevated temperature where specimens were preheated without load to target temperature (50 °C below softening point) and then soaked for one hour before testing at 20 MPa min^{-1} to failure.

Specimens with an embedded thermocouple in the centre and one on the outside were heated at 5 °C min^{-1} to 800 °C. There was an initial expected lag between outer and inner temperatures which peaked at around 200 °C. The difference then started to decrease even becoming negative (centre hotter than outside) in 2 cases i.e. heat was evolved in the centre of the sample. This continued to around 300 °C. The highest degree of strength increase is associated with a high degree of heat evolution. The geopolymer with the lowest initial strength was hypothesised to have higher unreacted precursor levels which can convert to geopolymer at high temperature and therefore increase the strength.

Guerrieri and Sanjayan (2010) showed that higher initial geopolymer strengths were associated with lower ductility. These high strength samples have less capacity to accommodate thermal incompatibility due to temperature differences between the exterior and interior of specimens during heating. This was the reason for low strength retention values in these initially strong samples.

Kong et al. (2007) compared geopolymers based on metakaolin and fly ash at elevated temperatures. The strength of the fly ash geopolymer increased after firing at 800 °C whilst the metakaolin showed a decrease in strength. The metakaolin

5.2 Thermal Expansion

mix had a solids to liquid ratio of 0.8 whilst the fly ash was 3.0. The metakaolin geopolymer pores were predominantly mesopores (1.25–25 nm) whilst the fly ash geopolymer pores had a significant proportion of micropores which facilitate water escape at elevated temperatures thus causing minimal damage to the geopolymer structure. The strength increase in the fly ash based geopolymer is also attributed to sintering of unreacted fly ash remnants.

Kong and Sanjayan (2008) investigated the effect of commonly used coarse aggregates in geopolymers. Aggregates are usually added to geopolymer pastes at 70–80 vol% and therefore contribute markedly to the overall thermal expansion of the resulting composite. The presence of aggregate resulted in a reduction in compressive strength of the composite after exposure to elevated temperatures. This is attributed to the mismatch in thermal expansion between geopolymer paste and aggregate. Figure 5.5 shows the thermal expansion of some coarse aggregates.

Kong and Sanjayan (2010) produced fly ash based geopolymer binders using sodium silicate and potassium hydroxide. Coarse aggregates based on crushed basalt and steel mill slag were used to produce concretes. The fine aggregate was river sand with 3 mm maximum size. This work compared geopolymer paste (no added aggregate), mortar (added sand only) and concrete (using sand plus aggregates ranging from 2.37 to 20 mm).

The 3 day compressive strengths of paste, mortar and concrete were similar, but after heating the paste had a 73.4 % strength loss, the mortar showed 100 % strength loss and the concrete 58.4 % strength loss.

Three sizes of compression strength sample based on paste were produced, cubes $25 \times 25 \times 25$ mm^3, and cylinders 70 mm high $\times$ 35 mm diameter and 200 mm high $\times$ 100 mm diameter.

Exposure to elevated temperature affected strength changes to larger specimens more adversely than smaller specimens. The cube specimens showed a small gain of 6.4 % whilst the cylinders showed strength losses of 52.3 and 73.4 % with increasing cylinder dimensions. The variation in strength loss was attributed to larger thermal gradients in larger specimens which led to thermally induced cracking.

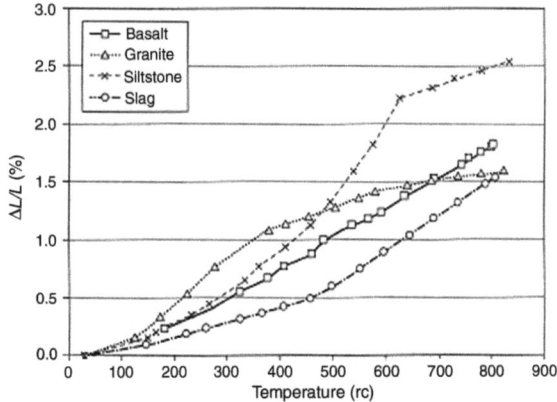

Fig. 5.5 Thermal expansion of coarse aggregates (Kong and Sanjayan 2008)

The addition of coarse aggregate to form geopolymer concrete showed that aggregate size is important to thermal stability. Concretes made with aggregate size grading of 2.36–5 mm and 5–10 mm exhibited spalling at 420 and 505 °C, respectively. A consistent 61.8 % strength loss was recorded for the 10–14 mm and the 20 mm sized aggregates.

The aggregate which has a higher stiffness than geopolymer paste, restrains the matrix shrinkage. This restraining effect sets up radial and tangential stresses around the aggregate particles. The shrinkage arising from water loss enhances these stresses, leading to the development of extensive cracking. This behaviour is more evident in concretes made with smaller sized aggregate.

Fernandez-Jimenez et al. (2010) compared OPC paste with fly ash based geopolymer and a blend of 70 % fly ash and 30 % OPC activated with sodium carbonate and sodium silicate both in solid form. Physical tests were carried out in-situ at elevated temperature up to 600 °C and post firing after exposure in steps from 200 to 1,000 °C followed by cooling to ambient temperature.

At temperatures from 400 to 600 °C Portland cement strength declined drastically due to water loss and degradation of the C-S-H phase. This occurred in both test regimes. Once the material cooled it failed to regain any of the initial strength. The fly ash based geopolymer performed better than Portland cement with residual strength being maintained or showing some gains. The presence of alkali in the geopolymer favoured the formation of a molten phase at 600–700 °C. The solidification of this molten phase on cooling may explain the increased strength on cooling. The presence of this liquid phase could also account for the drop in strength at 600 °C in the in-situ test regime.

The fly ash—OPC blend showed intermediate behaviour change, with lower initial strength than the both pure systems, but performed better at high temperature than the OPC.

Research by Chen-Tan et al. (2009) had shown that only amorphous aluminosilicates in the fly ash will react to form geopolymers.

Rickard et al. (2011) characterised three Australian fly ashes in order to gain an understanding of the properties relevant to producing geopolymers with good elevated temperature resistance. Variation arises in fly ash due to coal composition and power station furnace operating conditions. The total composition was obtained by X-Ray fluorescence (XRF) and crystalline phase components by X-Ray diffraction (XRD). Loss on ignition (LOI) was determined up to 1,000 °C.

The amount of potentially reactive amorphous material is obtained by manipulation of XRF and quantitative XRD (QXRD) results. Geopolymers from the three fly ashes were produced at a range of silicon:aluminium ratios and then fired at 1,000 °C. Changes in strength after firing were recorded. In two of the samples sintering of unreacted fly ash material occurred at elevated temperature leading to strength increases of 200–400 %. The other fly ash geopolymer exhibited high strength prior to firing, but showed a marked decrease in strength (>70 %) after firing. This was attributed to oxidation of iron compounds which led to a thermal expansion with resulting crack formation. XRF showed that this latter fly ash had 13.2 wt% iron oxides. This particular fly ash would be unsuitable for the manufacture of geopolymers for high temperature applications.

5.2 Thermal Expansion

SEM showed that the initial low strength geopolymers contained partially reacted fly ash particles bonded by the geopolymer gel. For all samples the higher the silicon to aluminium ratio the higher the retention of strength after firing.

The following additional points were made:

- Finer fly ash particles are preferable for higher initial strength properties.
- A spherical morphology is more suitable for low water mixes without loss of workability. Lower water content is also beneficial for reduced shrinkage at elevated temperatures.
- The presence of quartz particles in the fly ash can reduce workability and may also lead to expansion cracking at elevated temperatures.

Rickard et al. (2012) continued the above work with a more extensive investigation of fly ashes from five Australian power stations. Geopolymers were synthesised from each fly ash at a range of silicon to aluminium ratios. Only the amorphous (glassy) components were considered as reactive in geopolymerisation.

The silicon to aluminium ratio in the glassy component of the fly ash influenced the thermal properties of the geopolymers. High silicon to aluminium ratios (>5) showed compressive strength gains and better dimensional stability after exposure to 1,000 °C, whilst low silicon to aluminium ratios (<2) exhibited loss of strength and decreased dimensional stability.

In this work dilatometry was performed across the range of geopolymer samples with silicon to aluminium ratios of 2.0, 2.5 and 3.0. Marked differences between individual fly ashes were observed and in some cases with changing silicon to aluminium ratio of the geopolymer made from the same fly ash.

Geopolymers made with added alkali silicate solutions to achieve the required silicon to aluminium ratios showed higher shrinkage levels and even major expansion events in two cases. This expansion was attributed to swelling of unreacted silicate phases present from the activating solution.

Using SEM a higher proportion of binding phase was observed in geopolymers produced by silicate activation which led to greater initial strength than sodium aluminate activated geopolymers. After firing geopolymer strengths were affected by variations in phase changes in the non geopolymer phases as well as damage due to dehydration. Sintering was the main factor leading to compressive strength increases as it released more aluminosilicate material from the unreacted fly ash and activating solution remnants into the binding phase and improved inter-particle connectivity. This effect was consistently evident across all the samples and as such the cause for variable post fired strength must lie elsewhere. It is thought that activating solutions with high silicate levels showed incomplete incorporation of the silica leaving residual silicate phases which lead to lower post fired compressive strength.

Whilst all the geopolymer samples exhibited strength improving microstructural changes such as improved particle-particle bonding after firing, the instability of the non geopolymer phases after firing led to strength reductions. Typically oxidation of iron oxide phases leading to thermal expansion and associated cracking and the expansion of added unreacted silicate phases can be the cause of strength losses after firing.

5.3 Thermal Conductivity

In solids the carrier of heat is the phonon, a quantum of lattice vibrational energy. Whilst the theoretical model for the mechanism of phonons in heat transfer is based on crystalline solids some of the basic ideas may be used to gain some understanding of thermal conductivity in other materials. The mean free path is the distance between phonon scattering centres. Heat is mostly carried by phonons in the acoustic range i.e. with frequencies less than 20 kHz.

$$k = \frac{1}{3}Cvl \tag{5.8}$$

$$v = \sqrt{(E/\rho)} \tag{5.9}$$

where

k = thermal conductivity of medium,
C = specific heat of medium,
v = velocity of the carrier of heat (phonons),
l = mean free path,
E = Young's modulus of medium,
ρ = density of medium.

Klemens and Gell (1998) reviewed the thermal conductivity of thermal barrier coatings based on ceramic materials. Heat is conducted by lattice waves and by a radiative component which becomes significant at elevated temperatures. Point defects such as added cations and lattice vacancies reduce the lattice thermal conductivity by scattering high frequency lattice waves (phonons). Grain boundaries scatter lattice waves in the low frequency part of the spectrum. Both effects can be enhanced by adding cations and decreasing average grain size. To reduce the radiative component larger imperfections around 1 μm in size are required. These imperfections could take the form of a heat stable solid or preferably pores.

Clarke (2002) reviewed the theories of thermal conductivity with reference to temperatures beyond the Debye temperature. He discussed the influence of pores on decreasing the conductivity of materials. The volume fraction of pores, their aspect ratio and spatial distribution all need to be considered. An illustration of this is the formation of flat "pancake" like pores perpendicular to the temperature gradient during plasma spray application of ceramic coatings which led to the greatest decrease in thermal conductivity. He concluded that a material will have low thermal conductivity at high temperatures if it satisfies:

- High molecular weight
- Complex crystal structure
- Non-directional bonding
- Large number of different atoms per molecule

5.3 Thermal Conductivity

Table 5.3 Calculated minimum thermal conductivities (W m^{-1} K^{-1}) (Clarke 2002)

Compound	K$_{min}$	Compound	K$_{min}$	Compound	K$_{min}$
BeO	3.78	MgAl$_2$O$_4$	2.34	NiO	1.48
SiC	3.00	TiO$_2$	2.07	LaMgAl$_{11}$O$_{19}$	1.48
Al$_2$O$_3$	2.89	Mg$_2$SiO$_4$	2.00	Gd$_2$Zr$_2$O$_7$	1.14
MgO	2.56	Mullite	1.68	Monazite	1.13
AlN	2.45	ZrO$_2$ (YSZ)	1.49	ThO$_2$	0.98

The data shown in Table 5.3 and Fig. 5.6 were used to select additives for alumina and zirconia based thermal coatings to reduce thermal conductivity levels.

5.3.1 Thermal Conductivity of Geopolymers

Thermal conductivity determination is required to assess geopolymers' suitability for potential applications in thermal barriers and construction structural members. Insulators require a low k value to reduce heat flow by conduction. Structural concrete requires higher k values to allow for a reduction in thermal gradients and resulting reduced expansive stresses (Subaer and van Riessen 2007). Subaer obtained a 40 % increase in thermal conductivity (0.65–0.91 W m^{-1} K^{-1}) by adding 40 wt% quartz aggregate to a metakaolin based geopolymer.

Liefke (1999) produced geopolymer foams with densities in the range 100–800 kg m^{-3} with a k value of 0.037 W m^{-1} K^{-1}. He also reported on open cell foams

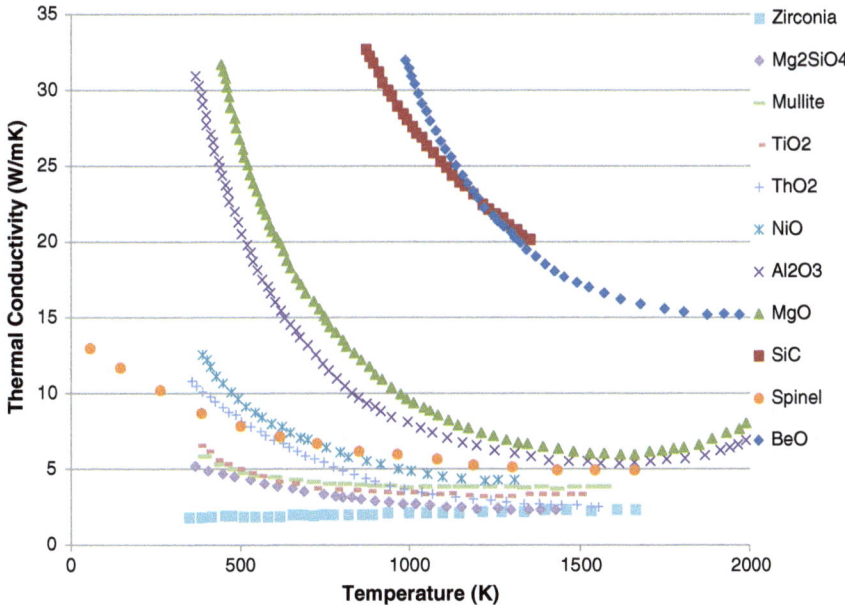

Fig. 5.6 Temperature dependence of thermal conductivity (Clarke 2002)

with densities in the range 350–400 kg m^{-3} and a k value of 0.16 W m^{-1} K^{-1}. This latter foam is suitable for automotive exhaust silencer components.

Rickard et al. (2013) produced metakaolin foams of various densities and measured k values. Foam with a density of 800 kg m^{-1} gave a k value of 0.3 W m^{-1} K^{-1}. The fire rating of this sample tested to AS1530.4 was 64 min for a 50 mm thick specimen.

Jonker et al. (2009) developed a geopolymer foam system suitable for refractory service. The material was based on a blend of fly ash and metakaolin (5:1 by weight) and activated with sodium silicate and sodium hydroxide. Powdered aluminium metal was the blowing agent. The density was 750 kg m^{-3} after drying at 110 °C and 600 kg m^{-3} after firing at 1,100 °C. Compressive strength was 4 MPa and increased to 14 MPa after firing at 1,100 °C. Fired shrinkage was less than 2 %. Thermal properties are shown in Table 5.4.

Porosity was the main influence on lowering thermal conductivity. The pores should be as small as possible and cracks and coarse pores of more than 5 mm should be avoided. The higher content of small pores is beneficial to decreasing thermal conductivity and increasing strength of the material.

The value of k is driven by different mechanisms in different materials. Metals have high electrical and thermal conductivity. Metals with a face centred cubic structure e.g. silver (k = 430 W m^{-1} K^{-1}) have the highest electrical and thermal conductivities, whilst body centred cubic metals are an order of magnitude lower. Ceramics show low conductivity due to the large band gap between valence and conduction bands (Rollett 2007).

Duxson et al. (2006) used thermal conductivity measurements to investigate the microstructure of metakaolin based geopolymers up to 100 °C. Water will still be present up to 100 °C and this can have a masking effect on measurements. Increasing silicon to aluminium ratios between 1.15 and 2.15 gave increased k values with sodium activated geopolymers giving slightly higher values than potassium activated geopolymers. Over the temperature range investigated k values showed minor changes.

Kamesu et al. (2012a) produced geopolymers using metakaolin from two sources, one of which had high quartz content, and measured the following thermal properties, thermal diffusivity, thermal resistance and thermal conductivity. Thermal diffusivity was determined by the laser flash method and the thermal conductivity determined from Eq. 5.5.

Table 5.4 Thermo-physical properties of geopolymer insulating material (Jonker et al. 2009)

Temperature (°C)	Thermal diffusivity (mm^2 s^{-1})	Specific heat (J g^{-1} K^{-1})	Thermal conductivity (W m^{-1} K^{-1})
23	0.378	0.740	0.227
199	0.348	0.935	0.264
401	0.350	1.037	0.293
601	0.366	1.098	0.326
800	0.387	1.171	0.368
998	0.416	1.230	0.415
1,098	0.447	1.273	0.462

5.3 Thermal Conductivity

Porosity determination and water contents were measured and related to thermal properties. As the silicon to aluminium ratio of the geopolymers increased thermal diffusivity increased. At Si:Al = 1.23, the thermal diffusivity was 2.24×10^{-7} m^2 s^{-1} and increased to 3.83×10^{-7} m^2 s^{-1} at Si:Al = 2.42. These results equate to thermal conductivity values of 0.3–0.59 W m^{-1} K^{-1}.

Increasing the silicon to aluminium ratio leads to a decrease in cumulative pore volume with a corresponding reduction in average pore size. The thermal properties are related to porosity and pore interconnectivity of the geopolymer. The following mechanisms were suggested for the change in humidity of geopolymer specimens:

1. Chemical fixing of atmospheric water which is more predominant in the case of higher Si:Al ratio specimens. This is via reaction of water with silicon sites to form Si(OH)$_4$. For samples with low Si:Al molar ratio it takes longer for the humidity to be fixed due to a lower concentration of active silica sites.
2. Physical absorption of atmospheric water in the pores.

The authors concluded that thermal conductivity for insulation applications can be influenced by Si:Al molar ratio, alkali content, and crystalline phase content.

Kamesu et al. (2012b) produced geopolymer foams from calcined silica rich kaolin and presented two theoretical models for the variation of effective thermal conductivity with total porosity. The variation in Si:Al ratio in these formulations also includes the residual quartz which is not transformed in the kaolin calcination step. Residual quartz will increase the effective thermal conductivity of the geopolymer since crystalline quartz has k values in the range of 6–11 W m^{-1} K^{-1} whilst amorphous silica is three to five times lower.

With the increase in total porosity, the effective thermal conductivity decreases. This is illustrated in Figure 5.7. The addition of blowing agent (aluminium powder) changes the pore size, volume fraction and spatial arrangement of the pores.

Khan (2002) measured thermal conductivity on a range of aggregates alone and in OPC concrete. Moisture played an important role on the measured

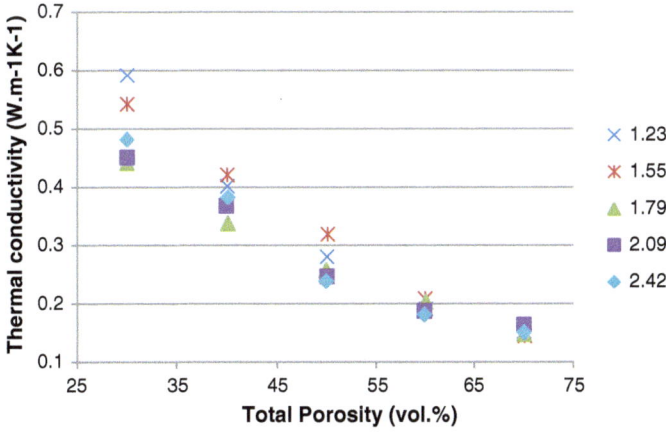

Fig. 5.7 The influence of total porosity on thermal conductivity (Kamseu et al. 2012b)

Table 5.5 Thermal conductivity of aggregates (Khan 2002)

Type of aggregate	Thermal conductivity (W m^{-1} K^{-1})		Specific gravity
	Dry state	Wet state	
Basalt	4.03	4.30	2.70
Limestone	3.15	3.49	2.69
Siltstone	3.52	5.22	2.66
Quartzite	8.58	8.63	2.67

Table 5.6 Thermal conductivity of mortar and concrete in dry and fully saturated state (Khan 2002)

Type of concrete	Thermal conductivity (W m^{-1} K^{-1})			
	SAND I		SAND II	
	Dry	Fully saturated	Dry	Fully saturated
Mortar	1.90	2.65	1.37	1.95
Basalt concrete	2.26	3.52	1.97	3.24
Limestone concrete	2.03	2.92	1.60	2.71
Siltstone concrete	2.21	3.61	1.91	2.90
Quartzite concrete	2.77	4.18	2.29	3.49

thermal conductivity value. Tables 5.5 and 5.6 show the variation of aggregate and concrete thermal conductivity values due to water content. Sand I is a quarried sand whilst Sand II is river sand containing mica. The higher thermal conductivity values for Sand I were attributed to the higher quartz content.

Chapter 6
Fire Resistance of OPC and Geopolymers

Abstract Geopolymer based systems have inherently superior fire resistance compared to Portland cement based and organic polymer systems. Geopolymer systems are substantially inorganic based and are considered incombustible, emitting no toxic fumes when exposed to fire. Compared to Portland cement based systems geopolymers retain a significant level of structural stability after exposure to fire events and show little if any explosive spalling. Spalling may be controlled by the addition of organic fibres which vaporise leaving a network of channels which facilitate water escape. Standard fire testing curves for the evaluation of cementitious materials are described together with the outcomes of using these test standards with both geopolymer and OPC components. The replacement of organic binders by geopolymer in woven fabric reinforced composites lead to systems meeting the Federal Aviation Authority (FAA) requirements. An extension of this work has lead to the development of lightweight, fire resistant coatings. A brief overview of passive fire systems for tunnels is included.

Keywords Spalling · Standard fire test curve · Tunnel linings · Fire resistant coatings and laminates

The influence of fire on materials is governed mainly by the composition of the material exposed with the following properties being the key to determining the level of fire resistance:

- Combustibility;
- Thermal conductivity;
- Resistance to thermal shock;
- Melting temperature;
- Structural stability (retention of mechanical properties);
- Inorganic/organic component ratio;
- Phase changes e.g. formation of chars, thermal expansion/shrinkage events, melting points and other transitions;

- Presence of pores which affects thermal insulation and vapour transportation;
- Presence of volatiles e.g. water;
- Propensity to spalling.

Geopolymer based systems have inherently superior fire resistance compared to Portland cement based and organic polymer systems. The fire resistance of cementitious systems is controlled by a range of factors. Geopolymer systems are substantially inorganic based and are considered incombustible, emitting no toxic fumes when exposed to fire. Compared to Portland cement based systems geopolymers retain a significant level of structural stability after exposure to fire events and show little if any explosive spalling. This lack of explosive spalling in geopolymers is attributed to an interconnected pore system which allows easy passage of volatiles, predominantly water, through the geopolymer structure when a thermal gradient is applied. The majority of the water in geopolymers is not bound in hydrates as is the case with Portland cement based systems. In the latter case the loss of water from the hydrates present results in the reduction of strength and consequent load bearing ability (van Riessen et al. 2009; Kong et al. 2007).

Resistance of geopolymers to fire exposure is managed by the techniques used to control thermal expansion/shrinkage events and loss of volatiles, predominantly water, to leave a post fire material in a serviceable condition.

The fire supplies the thermal energy which drives the changes to the exposed material. A fire requires a fuel source, an oxidising agent and an ignition source. The reaction is exothermic in nature and proceeds via a free radical mechanism (Higgins 2008).

One of the earliest attempts to investigate the combustion reaction was by Michael Faraday (1861) who presented his findings in a series of lectures for juveniles at the Royal Society in 1860/1861. These six lectures were published as "The Chemical History of a Candle" and are still in print to this day.

6.1 Fire Testing

The development of standard fire testing regimes needs to account for the wide range of conditions which may occur. In addition to the actual exposure temperatures occurring the rate of temperature increase needs accommodating in the test plan. Opinion of what should constitute a "standard fire" has led to considerable discussion. van Riessen et al. (2009) showed a temperature versus time curve for a typical room fire (Fig. 6.1) which consisted of three stages:

1. The growth or pre-flashover stage where the average temperature is low and the fire is localised in the vicinity of its origin;
2. The fully developed or post-flashover fire. All the combustible materials in the room are involved and flames appear to fill the entire room volume;
3. Decay or cooling period.

Fig. 6.1 Typical room fire curve (Institution of Engineers Australia 1989)

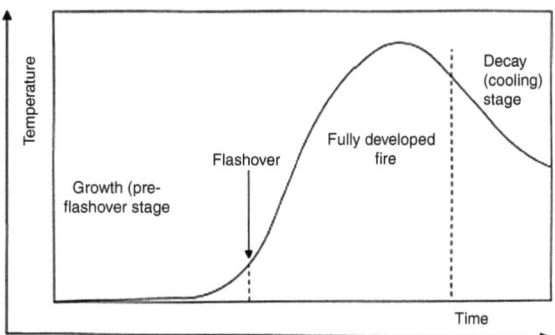

Flashover is a phenomenon unique to compartment fires, where products of incomplete combustion collect at the ceiling and then ignite leading to total involvement of the compartment materials and signalling the end of human survivability. The time to flashover is the available escape time. The Federal Aviation Administration (FAA) has used time to flashover of materials in aircraft cabin tests as the basis for acceptance criteria for commercial aircraft cabin materials (Lyon et al. 1997).

In the case of geopolymers no incomplete combustion products are formed during a fire so flashover times are theoretically infinite. Organic polymers used as binders for fibre reinforcement in composites generate smoke (incomplete combustion products), which lead to flash over times of 10–25 min. Fibre content of conventional composites is in the range 20–50 wt%, whilst advanced composites can have fibre contents as high as 70 wt% (Budinski and Budinski 2005).

Lawson (2009) reviewed the history of fire testing. The statement of the first law of thermodynamics in the nineteenth century led to the development of tools for calorimetry. In the early twentieth century the global recognition that scientific based fire standards were needed to protect lives and assets was realised. In the latter part of the twentieth century the development of data recording and analysis has led to more detailed studies of fire behaviour.

Ongoing developments based on the above have led to improvements in the testing of structural components, assemblies and systems under realistic fire and loading conditions. This should lead to enhanced structural performance to ensure building fires do not lead to partial or complete collapse scenarios. In addition improvements in ignition resistance, surface flammability, fire growth and smoke and toxic gas evolution of materials has led to the use of safer materials of construction.

The use of standard fire curves (temperature vs. time) to measure the fire resistance of and compare a range of materials is mandatory before materials may be used in construction. The scale of the testing (laboratory to full scale) also needs to be considered when evaluating the results of fire tests. The fire curves shown in Fig. 6.5 are for cellulosic fuel supply (AS1530.4/ISO 834 and ASTM E119) and hydrocarbon fuel source (Eurocode 1991-1-2). The cellulosic curve is based on the

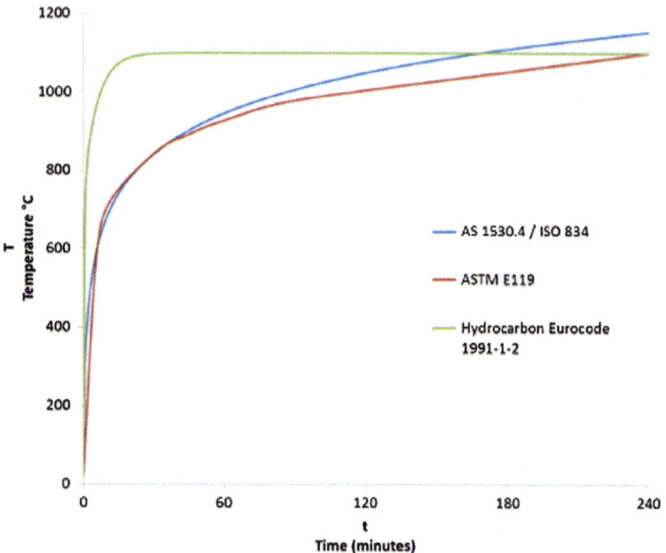

Fig. 6.2 Temperature versus time curves for standard fires (van Riessen et al. 2009)

fire curve in Fig. 6.2 and is used to replicate residential and commercial building fire scenarios. The hydrocarbon curve shows a greater rate of temperature increase and is used to replicate fires occurring in tunnels, chemical plant and oil and gas production and refining.

The Cement, Concrete and Aggregates Association of Australia issued report TC 61 (2010).

It listed three relevant concrete failure criteria:

1. Structural adequacy (ability to resist load);
2. Integrity (ability to resist the passage of flames);
3. Insulation (the ability to prevent fire spread due to unacceptable temperature rise of the unheated face).

In the same reference the importance of real fires (as opposed to standard fire curves) is emphasised. The rate of temperature rise is dictated by the rate at which fuel is pyrolised i.e. converted from solid to combustible gases, and then burnt when these gases come into contact with air. Once burning occurs there is an increase in pyrolisation due to heating. Given more air additional heat is generated. A test conducted in a compartment with a toughened glass façade found that glass breakage occurred after 18 min and gave rise to temperatures in the order of 1,000 °C. The final increase of air temperature from 300 to 1,000 °C took 5 min. AS 1530.4 requires a temperature of 841 °C after 30 min and to reach 1,006 °C after 90 min. Real fires can achieve conditions outside of those specified and these extreme events need to be factored into fire safety calculations.

6.1 Fire Testing

Several fire curves have been developed for specific applications. The RWS curve was developed by Rijkswaterstaat, Ministry of Transport in the Netherlands. This curve is based on a worst case scenario that a 50 m^3 fuel, oil or petrol tanker fire with a fire load of 300 MW could occur, lasting up to 120 min. The RWS curve was based on results of testing carried out by TNO in 1979. This is the basis for the Hydrocarbon curve (HC).

The RABT curve was developed in Germany in 1994 as a result of a series of test programmes. In the RABT curve the temperature rise is very rapid up to 1,200 °C within five minutes. The duration of the 1,200 °C exposure is shorter than other curves with the temperature drop off beginning at 30 min for car fires and 60 min for train fires. The 110 min cooling period is the same for both cases.

The HCM (hydrocarbon modified) curve is a French regulation with a maximum exposure temperature of 1,300 °C compared to the 1,100 °C of the standard HC curve.

The initial temperature gradient of all these curves is severe possibly causing thermal shock to the exposed concrete with resultant spalling.

Alarcon-Ruiz et al. (2005) used thermal analysis to follow the irreversible decomposition reactions taking place as cured cement paste was fired in 100 °C steps up to 800 °C. The thermal decomposition of the cement paste was studied by thermogravimetric analysis (TGA) and the derivative thermogravimetric (DTG) curves. Such techniques should be able to determine fire conditions and the resultant deterioration of the cement paste. These techniques can be used to assess the state of concrete after fire exposure. The TGA work showed that the dehydration and decarbonation reactions are irreversible in the cement paste and can be used as tracers in determining temperature history after fire exposure. The dehydroxylation reaction of portlandite is reversible with the reaction occurring rapidly after heat treatment.

6.2 Portland Cement

Denoël (2007) reviewed fire safety and concrete structures for the Federation of the Belgian Cement Industry. This covered the choice of concrete, regulations, fire resistance, fire risk, fire safety engineering and repair of concrete after fire.

Hertz (2003) investigated spalling in traditional and high strength concrete back to the mid 1980s and gives a comprehensive list of possible causes of spalling. Several other authors (Papworth 2000; Kalifa et al. 2001; Peng et al. 2006) have carried out fire testing of high strength concrete.

Papworth (2000) tested steel reinforced slabs 4.5 m × 1.1 m × 0.250 m thick containing silica fume and nylon fibres which he considered gave advantages over polypropylene. The fire curve used was based on AS 1530.4 (Standards Australia 2005). The slab containing the nylon fibre and silica fume showed some fine cracks on the fire side, but no spalling. Three other samples, two with silica fume only and a control (no silica fume) showed significant spalling. The slabs were cured for 28 days prior to testing to ensure moisture was still present in the slabs. This was an attempt to replicate a tunnel lining where the buried inner concrete surface has access to ground moisture.

Kalifa et al. (2001) added polypropylene fibres to high strength concrete to reduce spalling. Pore pressure measurements performed on heated concrete showed the presence of fibres lead to a large decrease in the extent of pressure zones which built up in the porous network. Fibres were found to have only a small effect on the thermal properties of concrete. In the fire test using the ISO 834 fire curve, the temperature rise shows a plateau effect that starts above 100 °C and ends between 160 and 220 °C, which is in the melting range of the fibres. In plain concrete the plateau ends at around 250 °C. This phenomenon is associated with the consumption of energy due to water vaporisation in the porous network. The end point of this phenomenon correlated with the measured pressure peak. The melted fibres enable vaporised water to be expelled from the concrete at lower temperatures. Pressure versus time curves have a similar shape at all fibre dosages, but the pressure peak height is drastically reduced with increasing fibre dosage.

Peng et al. (2006) added hybrid fibre blends, based on polypropylene and steel, to high strength concrete and obtained improvements in fracture energy which could be correlated to improved spalling resistance. Exposure to 400 °C showed a drop in residual strength, but fracture energy was higher than at ambient temperature only in the samples containing steel fibres. An increase in fibre pull out values was observed in samples heated to 400 °C. Above 400 °C strength reduced rapidly in all specimens. Peng concluded that a hybrid fibre blend using polypropylene fibres to form a pathway for water vapour escape after melting and a steel fibre to inhibit crack propagation during heating would be beneficial to heat resistance. Spalling was observed in some of the samples containing polypropylene fibre and this was attributed to the low loading (0.25 kg m^{-3}) and the dense structure of the concrete made at a w/c of 0.26. He suggested increasing polypropylene dosage to 2 kg m^{-3} to improve spalling resistance.

Both (2003) considered there were three options to improve spalling resistance of concrete:

1. Increase permeability preferably during the duration of the fire to avoid durability issues;
2. Increase fracture energy of the concrete;
3. Reduce difference in coefficient of thermal expansion between matrix and aggregates.

He considered low melting point fibres to address option 1. Two types of polypropylene fibres were compared at 2 kg m^{-3} in a standard tunnel lining concrete mix. The first was a monofilament, 12 mm long × 18 μm diameter and the second was a fibrillated fibre also 12 mm long, but 60 μm diameter. Fibrillated fibres have been processed to give a branched structure with increased surface area compared to the parent monofilament fibre. 0.35 m thick concrete slabs with 4 wt% water were exposed to the RWS fire curve. Test results showed that the samples containing monofilament were only superficially damaged whilst the fibrillated samples showed spalling to a depth of 35 mm.

Based on the above work three levels (1, 2, and 3 kg m^{-3}) of polypropylene monofilament were compared to a fibre free control in the standard tunnel mix. The fire curve applied in this test was the RABT ZTV curve, but with the 1,200 °C plateau extended out to 120 min. Spalling depths decreased from an average of 95 mm (265 mm maximum) for the fibre free control to 7 mm average (25 mm maximum) for the 3 kg m^{-3} mix.

To address option 2 the use of a hybrid blend of steel fibres (50 kg m^{-3}) and fibrillated polypropylene (0.9 kg m^{-3}) in a 50 MPa compressive strength concrete mix was evaluated using the RABT ZTV curve with the heating plateau extended to 120 min. A complex set of stresses is set up during heating when expansion is restricted. The face exposed to the heat source is in compression, but with tensile forces further into the body of the concrete. The steel fibres can control these tensile stresses and reduce the spalling tendencies. An average spalling depth of 30 mm was obtained with a standard deviation of 20 mm.

Bilodeau et al. (2004) investigated the resistance of high strength concrete, filled with expanded slate as part of the aggregate, to exposure to a hydrocarbon fire curve. Two w/c ratios, 0.33 and 0.42 were evaluated together with the addition of two types of polypropylene fibre at a range of dosages. The fibres were described as fibrillated with a length of 20 mm and as a multifilament, 12.5 mm long and significantly thinner than the fibrillated fibre. He concluded that there were more fibres per unit volume of the thinner fibre and that this contributed to the improved spalling resistance.

The thinner 12.5 mm long fibre was more efficient than the 20 mm fibre in preventing spalling. 1.5 kg m^{-3} of the thinner fibre was required compared to 3.5 kg m^{-3} of fibrillated fibres. This is in line with the work of Both.

6.3 Fire Resistance of Geopolymers

Davidovits' original impetus with geopolymer research was driven by the development of fire resistant systems after several tragic fires in Europe in the 1970s (Davidovits 1989). Davidovits (2008d) used a 10 mm thick geopolymer panel which was exposed to a 1,000 °C flame and measured the cold side temperature. After 30 min the values shown in Table 6.1 were obtained.

The dehydroxylation heat (endothermic) of chemically bound water and hydroxyl groups varies with geopolymer type (and Si:Al ratio) and the higher ΔH values can explain the lower cold side temperatures. More heat is required for the water release mechanism to occur therefore less is available to raise the sample temperature. In Fig. 6.3 a plateau exists between 4 and 12 min for the Na-poly(sialate) sample which is related to the higher enthalpy demand for this sample.

Table 6.1 Temperature and heat evolution during fire testing (Davidovits 2008d)

Geopolymer type	30 min temp. (°C)	Si:Al	ΔH kcal/mole water	ΔH cal/mole GP
Na-poly(sialate)	180	1	30	300–500
K-poly(sialate-siloxo)	270	2	22	100–200
K-poly(sialate-disiloxo)	300	3	20	50–100

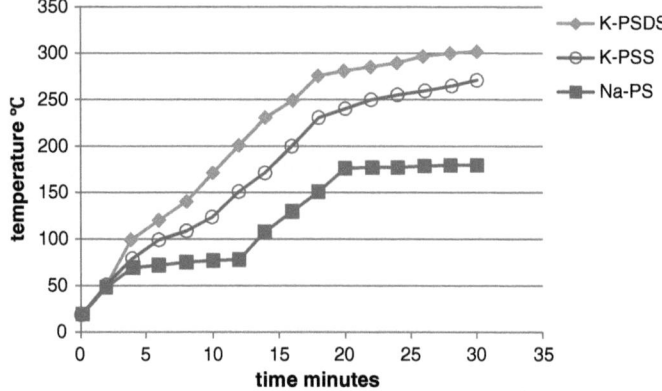

Fig. 6.3 Fire testing results for metakaolin based geopolymers (Davidovits 2008d)

Lyon working in conjunction with Davidovits used geopolymers as binders for cross woven fabric laminates for aircraft applications (Lyon et al. 1997; Lyon 1996, 1999). Organic binders based on epoxy, vinyl ester, phenolic, advanced thermosets and high temperature engineering plastics were reinforced with either woven glass or carbon fibre fabrics and compared to carbon fibre reinforced geopolymer. The geopolymer outperformed every other system in fire testing in terms of weight loss, time to ignition, smoke, total heat release and flashover.

Fire resistant coatings based on geopolymers have been reported in the literature (Temuujin et al. 2010, 2011, 2012; Chen et al. 2011a). Giancaspro et al. (2006) applied a thin coating of geopolymer containing glass microspheres to balsa wood sandwich panels to act as a fire resistant barrier. A 1.8 mm thick coating satisfied the requirements of the Federal Aviation Administration (FAA) for both heat release and smoke generation.

Palomo and Fernandez-Jimenez (2011) investigated the use of fly ash based geopolymers for coatings to improve the fire resistance of fibre reinforced panels. Cement powder was used as a calcium source to obtain ambient cure. Coatings were 3–5 mm thick and the addition of 10 wt% filler increased fire resistance from 4 to 40 min. UNE-EN-ISO 11925-2 (International Organisation for Standardisation 2010) was the relevant standard used to obtain the fire resistance.

Won et al. (2012) investigated OPC/alkali activated blast furnace slag blends filled with ground porcelain for fire resistant systems suitable for lining of tunnels. The temperature curve used was the German RABT ZTV (van Alken 2012) curve which is suitable for testing linings for tunnels (Fig. 6.4).

6.3 Fire Resistance of Geopolymers

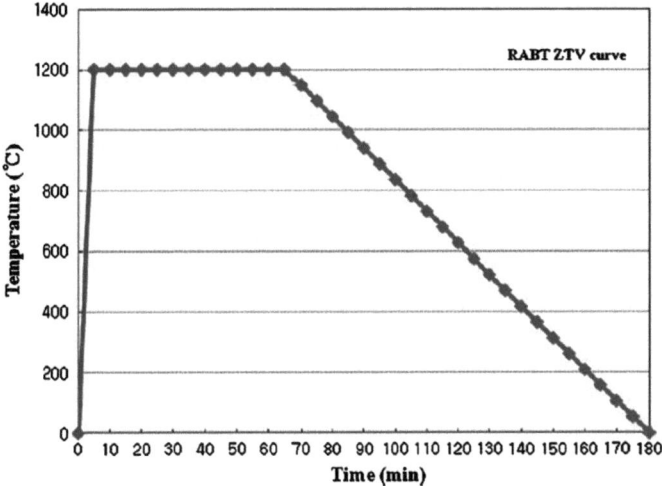

Fig. 6.4 RABT ZTV curve (Won et al. 2012)

Won measured residual compressive strength after fire testing. In the case of formulations with added porcelain (particle size = 4 to 6 µm) residual compressive strength increased as the amount of porcelain increased to a maximum of 30 wt% porcelain content. The effect of the porcelain was attributed to the formation of a vitreous phase by reaction of porcelain with silicate and alkali metal ions and resultant densification. Spalling occurred in some samples but this was alleviated by the addition of fibrillated polypropylene fibres (36 µm diameter and 6 mm long).

Zhao and Sanjayan (2012) compared OPC and fly ash based geopolymers at the same strength levels (40–100 MPa compressive strength) with regards to spalling resistance. They elucidated three possible mechanisms for spalling resistance:

1. Moisture clog spalling caused by steam pressure build up in the pores of the concrete in fire situations. When heated the water will begin to vaporise and pore pressure will increase. The vapour will migrate along the temperature gradient and either escape or condense if it reaches a sufficiently cool area in the concrete. This process continues forming a fully saturated layer. This saturated layer inhibits further migration of pore water. If vaporised water cannot escape quickly enough the internal pore pressure will continue to rise until it exceeds the tensile strength of the concrete and spalling occurs;
2. Spalling results from restrained thermal dilation close to the heated surface which, leads to compressive stresses parallel to the heated surface leading to brittle fracture of the concrete;
3. Thermal expansion incompatibility between concrete paste and aggregates, particularly siliceous types, can cause spalling.

Zhao and Sanjayan (2012) state that the probable cause of spalling is a combination of the presence of water and rapid heating rates. They also developed a test method to investigate spalling which applied a fire curve similar to a hydrocarbon fire to concrete cylinders. No spalling occurred in any of the geopolymer samples, but the high strength OPC samples containing silica fume showed severe spalling. Normal strength OPC showed minor spalling. Sorptivity, the ability to absorb and transmit water through the matrix by capillary suction was determined using ASTM C 1585. Sorptivity is a measure of the openness of the concrete pore structure. Lower sorptivity results indicated poorer spalling resistance.

Cheng (Cheng and Chiu 2003; Cheng 2003) investigated the production of fire resistant materials from blast furnace slag and waste serpentine mineral. Varying levels of potassium hydroxide and sodium silicate additions to slag/metakaolin blends were evaluated. Fire testing was carried out on 10 mm thick panels which were exposed to an 1,100 °C flame and the cold side temperature tracked against time. Increasing potassium hydroxide concentration reduced the cold side temperature of the sample (Fig. 6.5).

Increasing metakaolin levels had a similar but smaller effect on cold side temperature, whilst increasing sodium silicate addition showed little effect.

Cheng (2003) blended waste serpentine with metakaolin which was then activated with potassium hydroxide and sodium silicate. 10 mm thick panels were tested as above and compared to commercially available calcium silicate board. Both samples reached around 400 °C, but the geopolymer ramped up slower with a 4–6 min plateau at 100 °C.

The most important application for fire resistant concrete is tunnel linings. These can be in the form of precast segments, sprayed shotcrete and other coatings. Kim et al. (2010) reviewed the current scenario with tunnel linings and introduced a proposed coating for new and remedial concrete works. When tunnels are exposed to fire they can undergo severe damage and partial collapse. The high

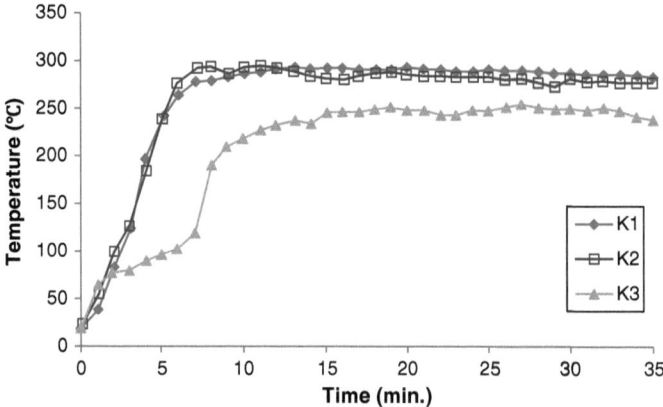

Fig. 6.5 Effect of KOH concentration on cold side temperature. K1 = 5 M; K2 = 10 M; K3 = 15 M (Cheng and Chiu 2003)

6.3 Fire Resistance of Geopolymers

relative humidity in tunnels (75 % +) can lead to explosive spalling due to water in the porous tunnel lining material. Tunnel linings are tested using the more stringent hydrocarbon fire curve. Loss of strength of tunnel linings can bring about long term loss of accessibility to the tunnel infrastructure with large economic losses.

Kim et al. (2010) lists four types of tunnel fire protection:

1. Start with a fire resistant shotcrete material as the main lining;
2. Use a coating material that is able to prolong heat transfer;
3. Spray a secondary lining onto the tunnel surface;
4. Install precast segments or panels.

A major focus has been on developing coatings using light weight or porous aggregates. However, properties of these systems tend to be low with compressive strengths typically less than 10 MPa. Pressures on tunnel linings due to typical road and rail traffic are in the range of 25–600 Pa. High speed train services produce higher pressures and induce vibrations in the lining. These high pressures and vibrations can cause negative pressures on the tunnel lining and result in spalling and fatigue failure. The use of fibres has improved the spalling resistance but can lead to lower strength linings.

The lining developed by Kim is based on OPC with bottom fly ash as the filler and 0.25 vol% polypropylene fibre. This material is a pre-mix type suitable for casting and shotcreting. The properties obtained are more than double those of currently used linings. Fire testing was carried out using the RABT curve designed to replicate tunnel fire conditions typical of hydrocarbon fires. 20, 30, and 40 mm thick coatings of the new system were applied over a sample slab of typical tunnel structure containing steel reinforcement and allowed to cure unsealed for 28 days at ambient. Whilst the 20 mm thick coating showed spalling, the other two thicknesses did not.

Tarada and King (2009) concluded that polypropylene fibres could reduce explosive spalling during cellulosic and hydrocarbon fires but were inadequate for the RWS fire curve. In the latter case the polypropylene could be augmented by panels and/or coatings. In the aftermath of a fire concrete assessment should be carried out in areas where the fibres have melted. This assessment should include as a minimum, concrete strength reduction and load bearing capability, permeability of the remaining concrete and influence on ongoing durability and risk of further spalling.

The panel and/or coating scenario must be designed so that their interface with the structural concrete does not exceed 350 °C in the event of a fire which will prevent spalling and significant changes in the structural concrete mechanical properties. Spray-on coatings have a one off use in a fire event and require replacement after such events. Typical panel thicknesses are 20–30 mm and coating applications are up to 50 mm.

Geopolymer based cellular materials (Rickard et al. 2013) have been evaluated for fire resistance. These metakaolin based geopolymers used aluminium metal to generate hydrogen gas as the blowing agent during processing. Fire ratings in

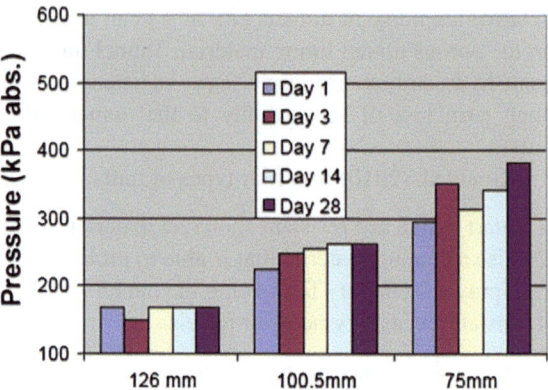

Fig. 6.6 Pore pressures estimated from the boiling points of water at different depths from hot face in concrete slab (Provis 2010)

excess of 60 min were obtained for 50 mm thick panels using the AS 1530.4 fire curve. Polypropylene fibres were used to control the extent of foam growth during the expansion process.

CSENG Ventures (Reid 2011) have a patent for fire resistant door and window frames where a hollow plastic, typically uPVC, section is filled with metakaolin based geopolymer foam. The geopolymer foam has sufficient structural stability in the event of a fire to seal off the fire compartment and prevent the ingress of air. Traditional wood and plastic frames burn and/or melt allowing air ingress to the fire.

Provis (2010) carried out fire testing on geopolymers based on blast furnace slag and fly ash using the ASTM E119 fire curve. The test was run for 4 h using 150 mm thick samples and the maximum cold side temperature recorded was 105.3 °C. An apparent flattening of the rate of temperature increase in the temperature-time curve occurred between 100 and 160 °C. This was associated with the passage of the boiling front through the slab. The author attempted to estimate the pore pressures from the boiling point of water, Fig. 6.6. The boiling point of water increases with pressure. Steam tables show the relevant pressures for individual temperatures. The plateau temperature value of the temperature-time curve corresponds to the boiling of water. By measuring the plateau temperature at different depths through the concrete specimen we can estimate the pressure at each depth. This is indicative of the passage of the boiling front. The curve flattening close to the hot face is not always distinct due to the fast passage of the boiling front at these positions.

The columns in Fig. 6.6 are indicative of sealed curing times at ambient temperature. The samples were then unwrapped and left to age at ambient temperature for 56 days when fire testing occurred. The shorter the sealed curing time the lower amounts of water at the start of fire test. No spalling was observed in any of these tests.

Chapter 7
Conclusions

The current extent of geopolymer commercialisation in dollar terms is extremely difficult to estimate. This can be attributed to the confidential technical and commercial aspects governing the competitive world of business. Many companies claim to have commercially available products based on geopolymers, but no output volumes are available. Certainly the green environmental aspects of geopolymers compared to OPC based products are widely used to promote geopolymer based products (Concrete Institute of Queensland 2010; Achille 2010; Banh UK Ltd. 2010).

The short term future for geopolymer based products is factory manufactured precast components, where heat curing and an acceptable level of technical control are readily available. This will enable speciality products such as sewage and chemical contacting components and thermally resistant parts such as tunnel linings to gain a market presence based on technical superiority. The move to ambient curing of geopolymer systems generally involves the introduction of calcium based products which can adversely impact on the excellent durability profile of geopolymers.

The large variations in fly ash composition and its influence on the geopolymerisation process and end product properties is still a major hurdle for larger scale acceptance of geopolymer based products. The current national standards for fly ash do not consider its application as a geopolymer feedstock. Specific test methods to classify fly ashes for the geopolymerisation process should be a priority for the relevant fly ash producers and marketing bodies. This will ensure that the majority of fly ash can be safely utilised in commercial products instead of creating environmental hazards.

References

3M. (2003). *Nexel selector guide and specification sheet* (Vol. 30, pp. 167–170). Technical Data Sheet. Retrieved from http://www.3m.com/ceramics.
Achille, F. (2010). Geo-green crete. *2010 Concrete Sustainability Conference* (p. 16). National Ready Mixed Concrete Association.
ACI Committee E2-00. (2006). Reinforcement for concrete—materials and applications. *ACI Education Bulletin, E2-00*, 16. Retrieved from http://www.concrete.org.
ACI Committee E-107. (2007). Aggregates for concrete. *ACI Education Bulletin, E-107*, 29. Retrieved from http://www.concrete.org.
Adfil. (2010). *Adfil ignis*, (p. 1). Technical Data Sheet. Retrieved from http://www.adfil.co.uk.
Agullo, L., Toralles-Carbonari, B., Gettu, R., & Aguado, A. (1999). Fluidity of cement pastes with mineral admixtures and superplasticizer—a study based on the Marsh cone test. *Materials and Structures, 32*, 479–485.
Akers, S., Kaufmann, J., Lübben, J., & Schwitter, E. (2009). Reinforcement of concrete and shotcrete using bi-component polyolefin fibres. *Conference Paper, Concrete Solutions 09* (p. 9). 17–19 Sept., Sydney, Australia.
Alarcon-Ruiza, L., Platret, G., Massieu, E., & Ehrlacher, A. (2005). The use of thermal analysis in assessing the effect of temperature on a cement paste. *Cement and Concrete Research, 35*, 609–613.
Alonso, S., & Palomo, A. (2001). Alkaline activation of metakaolin and calcium hydroxide mixtures: Influence of temperature, activator concentration and solids ratio. *Materials Letters, 47*, 55–62.
Alzeer, M., & Mackenzie, K. J. D. (2012). Synthesis and mechanical properties of new fibre-reinforced composites of inorganic polymers with natural wool fibr. *Journal of Materials and Science, 47*, 6958–6965.
Anasys Thermal Methods Consultancy. (2012). *Introduction to thermomechanical analysis*. Retrieved from http://www.anasys.co.uk.
Applied Sciences Inc. (2001). Research and Development Company. Retrieved from http://www.apsci.com.
Applin, K. R. (1987). The diffusion of dissolved silica in dilute aqueous solution. *Geochima and Cosmochimica Acta, 51*, 2147–2151.
Arioz, O. (2007). Effects of elevated temperatures on properties of concrete. *Fire Safety Journal, 42*, 516–522.
ASTM Committee A01-05. (2011). A820/A820M-11 standard specification for steel fibers for fiber-reinforced concrete. *ASTM Standard*.
ASTM Committee C09. (2012a). C618-12 coal fly ash and raw or calcined natural pozzolan for use in concrete. *ASTM Standard*, 3.
ASTM Committee C09. (2012b). C 494M-12 standard specification for chemical admixtures for concrete. *ASTM Standard*, 10.

ASTM Committee C16.30. (2010a). C177-10 standard test method for steady-state heat flux measurements and thermal transmission properties by means of the guarded-hot-plate apparatus. *ASTM Standard*, 23.

ASTM Committee C16.30. (2010b). C518-10 standard test method for steady-state thermal transmission properties by means of the heat flow meter apparatus. *ASTM Standard*, 16.

ASTM Committee C27-40. (2008). ASTM C1666/C1666M-08 standard specification for alkali resistant (AR) glass fiber for GFRC and fiber-reinforced concrete and cement. *ASTM Standard*, 4.

ASTM Committee D01-36. (2012). D7357-07 (reapproved 2012) standard specification for cellulose fibers for fiber-reinforced concrete 1. *ASTM Standard*, 3.

ASTM Committee D13-58. (2010). D7508/D7508M-10 standard specification for polyolefin chopped strands for use in concrete 1. *ASTM Standard*, 4.

ASTM Committee D20.30. (2009). D5930—09 standard test method for thermal conductivity of plastics by means of a transient line-source technique. *ASTM Standard*.

ASTM Committee E37. (2010). E289-04 (reapproved 2010) standard test method for linear thermal expansion of rigid solids with interferometry. *ASTM Standard*, 9.

ASTM Committee E37. (2011). E228-11 standard test method for linear thermal expansion of solid materials with a push-rod dilatometer. *ASTM Standard*, 10.

ASTM Committee E37. (2012). E831-12 standard test method for linear thermal expansion of solid materials by thermomechanical analysis. *ASTM Standard*, 4.

ASTM Committee C09-42. (2010). C1116/C1116M-10a standard specification for fiber-reinforced concrete 1. *ASTM Standard*, 7.

Australian Museum. (2007). *Classification of igneous rocks* (p. 2). Retrieved from http://www.australianmuseum.net.au.

Babatunde, A. O. Z., & Zhao, Y. Q. (2007). Constructive approach towards water treatment works sludge management: An international review of beneficial reuses. *Critical Reviews in Environmental Science and Technology, 37*, 129–164.

Bakharev, T. (2005). Resistance of Geopolymer materials to acid attack. *Cement and Concrete Research, 35*, 1233–1246.

Bakharev, T. (2006). Thermal behaviour of Geopolymers prepared using class F fly ash and elevated temperature curing. *Cement and Concrete Research, 36*, 1134–1147.

Bakhareva, T., Sanjayan, J. G., & Cheng, Y. B. (2000). Effect of admixtures on properties of alkali-activated slag concrete. *Cement and Concrete Research, 30*, 1367–1374.

Banh UK Ltd. (2010). *banahCEM geopolymer cement system*, (p. 2). Technical Data Sheet. Retrieved from http://www.banahuk.co.uk.

Banthia, N. (1994). Carbon fibre cements: structure, performance, applications and research needs. In J. Daniel & S. P. Shah (Eds.), *Fiber reinforced concrete, developments and innovations* (Vol. SP-142, pp. 91–120). Farmington Hills: American Concrete Institute.

Banthia, N., & Gupta, R. (2006). Influence of polypropylene fiber geometry on plastic shrinkage cracking in concrete. *Cement and Concrete Research, 36*, 1263–1267.

Barbosa, V., & Mackenzie, K. (2003a). Synthesis and thermal behaviour of potassium sialate geopolymers. *Materials Letters, 57*, 1477–1482.

Barbosa, V. F. F., & Mackenzie, K. (2003b). Thermal behaviour of inorganic geopolymers and composites derived from sodium polysialate. *Materials Research Bulletin, 38*, 319–331.

Barbosa, V., Mackenzie, K., & Thaumaturgo, C. (2000a). Synthesis and characterisation of materials based on inorganic polymers of alumina and silica: sodium polysialate polymers. *International Journal of Inorganic Materials, 2*(2), 309–317.

Barbosa, V. F. F., Mackenzie, K. J. D., & Thaumaturgoa, C. (2000b). Synthesis and characterisation of materials based on inorganic polymers of alumina and silica: sodium polysialate polymers. *International Journal of Inorganic Materials, 2*(2000), 309–317.

Barlet-Gaudedard, V., Zusatz-Ayache, B., & Porcherie, O. (2010). Geopolymer composition and application in oilfield industry. *US Patent 7, 794, 537*, p. 18.

BASF. (2008). Mode of action for superplasticisers for cement based construction chemicals. *Technical leaflet*. Retrieved from http://www.construction-polymers.com.

References

Bassioni, G. (2010). The influence of cement composition on superplasticizers' efficiency. *International Journal of Engineering (IJE), 3*, 577–587.

Bauer, S. W. (2007). *The history of the ancient world* (pp. 4–5). New York: W.W. Norton and company Inc. ISBN: 13:978-0-393-059748.

Bayer Materials Science. (2007). *BMS wins new partner for it's attractive market sector* (p. 1). Retrieved August 21, 2007, from http://nanotechnology-now.com.

Bayer Materials Science. (2008). *Dispersion-low viscosity material* (p. 1). Retrieved from http://www.baytubes.com/technology_and_applications/low_viscosity.

Bell, J. L., & Kriven, W. M. (2009). Preparation of ceramic foam from metakaolin based geopolymer gels. In H.-T. Lin, K. Koumoto, W. M. Kriven, E. Garcia, I. E. Reimanis & D. P. Norton (Eds.), *Developments in strategic materials* (pp. 97–111). Westerville: Copyright 0 2009 The American Ceramic Society.

Bentur, A., Diamond, S., & Mindess, S. (1985). The microstructure of the steel fibre-cement interface. *Journal of Materials Science, 20*, 3610–3620.

Bernal, S., de Gutierrez, R., Delvasto, S., & Rodriguez, E. (2006). Performance of geopolymeric concrete reinforced with steel fibres. *Conference Paper. IIBCC 10th International Inorganic Bonded Fibre Composites Conference* (pp. 156–167). Sao Paulo, Brazil.

Bernal, S. A., Bejarano, J., Garzón, C., Mejía de Gutiérrez, R., Delvasto, S., & Rodríguez, E. D. (2012). Performance of refractory aluminosilicate particle/fiber-reinforced geopolymer composites. *Composites Part B: Engineering, 43*, 1919–1928.

Bernal, S. A., Rodriquez, E. D., Megia de Guiterrez, R., Provis, J. L., & Delvasto, S. (2012). Activation of metakaolin/slag blends using alkaline solutions based on chemically modified silica fume and rice husk ash. *Waste Biomass Valor, 3*, 99–108.

Bilodeau, A., Kodur, V. K. R., & Hoff, G. C. (2004). Optimization of the type and amount of polypropylene fibres for preventing the spalling of lightweight concrete subjected to hydrocarbon fire. *Cement and Concrete Composites, 26*, 163–174.

Blissett, R. S., & Rowson, N. A. (2012). A review of multi component utilisation of coal fly ash. *Fuel, 97*, 1–23.

Bondar, D., Lynsdale, C. J., Milestone, N. B., Hassani, N., & Ramezanianpour, A. A. (2010). Geopolymer cement from alkali-activated natural pozzolans: effect of addition of minerals. *Conference Paper, Second International Conference on Sustainable Construction Materials and Technologies* (p. 9). June 28–30, Anacona, Italy.

Borinago-Trevino, R., Pascual-Munoz, P., Castro-Fresno, D., & Del Coz-Diaz, J. J. (2012). Study of different grouting materials used invertical geothermal closed loop heat exchangers. *Applied Thermal Energy, 17*(10), 956–967.

Both, C. (2003). Tunnel Fire safety. *Heron, 48*, 3–16.

Brew, D. R. M., & Mackenzie, K. J. D. (2007). Geopolymer synthesis using silica fume and sodium aluminate. *Journal of Materials Science, 42*(11), 3990–3993.

Brugge, C. (2011). *Concrix, macrofiber bi-component and high performance* (p. 4) Technical Data Sheet. Retrieved from http://www.bruggcontec.com.

Buchwald, A., Vicent, M., Kriegel, R., Kaps, C., Monzó, M., & Barba, A. (2009). Geopolymeric binders with different fine fillers—phase transformations at high temperatures. *Applied Clay Science,46*, 190–195.

Buchwald, A. W., J. 2009. ASCEM cement technology: Alkali activated cement based on synthetic slag made from fly Ash. In C. Shi & X. Shen (Eds.), *1st Int. Conf. on Advances in Chemically-activated Materials (CAM'2010, China)*, (pp. 15–21). RILEM Publications S.A.R.L., ISBN: 978-2-35158-101-8

Buckeye Building Fibres. (2009). *Ultrafibre 500, secondary reinforcement for concrete*, (p. 2). Technical Data Sheet. Retrieved from http://www.ultrafiber.com.

Budinski, K. G., & Budinski, M. K. (2005). *Engineering materials: properties and selection* (8th ed., p. 205). New Jersey: Pearson Education Inc. ISBN: 0-13-183779-6.

C-Therm. (2010). C-therm TCI thermal conductivity analyser. *Product Bulletin, 6*. Retrieved from http://www.ctherm.com.

Catlow, C. R. A., George, A. R., & Freeman, C. M. (1996). Ab initio and molecular-mechanics studies of aluminosilicate fragments, and the origin of Lowenstein's rule. *Chemical Communications*, 1331–1332.

CCANZ. (2009). Fibre reinforced concrete. *Information Bulletin, IB39*, 19. Retrieved from http://www.cca.org.nz.

Cement and Concrete Institute. (2010). Fibre reinforced concrete. *Information Bulletin*, 6. Retrieved from http://www.cnci.org.za.

Cement Concrete and Aggregates Australia. (2008). Use of recycled aggregates in construction. *Executive report* (No. 25). Retrieved from http://www.ccaa.com.au.

Cement Concrete and Aggregates Australia. (2010). Fire safety of concrete buildings. *Technical report CCAA T61* (No. 37). Retrieved from http://www.ccaa.com.au.

Cement Concrete and Aggregates Australia. (2005). *Plastic shrinkage cracking*. Technical Data Sheet. Retrieved from http://www.ccaa.com.au.

Central Federal Lands Highway Division. (2008). Concrete defects and curing chemistry. *Technical report* (No. 16). Retrieved from http://www.cflhd.org.

Ceratech. (2011). *Firerok high temperature concrete*, (p. 3). Technical Data Sheet. Retrieved from http://www.ceratechinc.com.

Chanh, N. V. (2004). *Steel fibre reinforced concrete* (pp. 108–116). Vietnam: Faculty of Civil Engineering, Ho Chi Minh City University.

Chaudhary, D., & Liu, H. (2009). Influence of high temperature and high acidic conditions on geopolymeric composite material for steel pickling tanks. *Journal of Material Science, 44*, 4472–4481.

Chen-Tan, N. W., van Riessen, A., Ly, C. V., & Southam, D. C. (2009). Determining the reactivity of a fly ash for the production of a geopolymer. *Journal of the American Ceramic Society, 92*, 881–887.

Chen, L., Han, W., Zhen, L., Wei, T., & Xiao, C. (2011). Preparation and properties of alkali stimulated geopolymer and its application in thermal insulating coatings. *Advanced Materials Research, 233–235*, 2443–2446.

Chen, N., Liu, M., & Yang, J. (1992). Influence of preparative history on physio- chemical properties of sodium aluminate solutions. *Chinese Journal of Materials Science and Technology, 8*, 135–137.

Chen, P.-W., & Chung, D. D. L. (1993). Carbon fiber reinforced concrete as an electrical contact material for smart structures. *Smart Materials and Structures, 2*, 181–188.

Chen, S. J., Collins, F. G., Macleod, A. J. N., Pan, Z., Duan, W. H., & Wang, C. M. (2011). Review paper carbon nanotube–cement composites: A retrospect. *The IES Journal Part A: Civil and Structural Engineering, 4*, 254–265.

Cheng, T.-W. (2003). Fire resistant geopolymer produced by waste serpentine cutting. *Proceedings of the 7th International Symposium on East Asian Resources Recycling Technology* (p. 4).

Cheng, T. W., & Chiu, J. P. (2003). Fire-resistant geopolymer produced by granulated blast furnace slag. *Minerals Engineering, 16*, 205–210.

Chindaprasirt, P., & Rattanasak, U. (2010). Utilization of blended fluidized bed combustion (FBC) ash and pulverized coal combustion (PCC) fly ash in geopolymer. *Waste Management, 30*, 667–672.

Chomarat. (2009). C-GRID. Retrieved from http://www.chomarat.com.

Chung, D. D. L. (1992). Carbon fibre reinforced concrete. *Report SHRP.ID/UFR-92-605*. Strategic Highway Research Program (No. 92).

Chung, D. D. L. (2005). Dispersion of short fibers in cement. *Journal of Materials in Civil Engineering, 17*, 379–383.

Clarke, D. R. (2002). Materials selection guidelines for low thermal conductivity thermal barrier coatings. *Surface and Coatings Technology, 163–164*, 67–74.

COI Ceramics. (2006). *Sylramic SiC fiber*, (p. 2). Technical Data Sheet. Retrieved from http://www.coiceramics.com.

References

Collepardi, M. (2000). Ettringite formation and sulphate attack on concrete. *Conference Paper, 5th CANMET/ACI International Conference on Durability of Concrete*, (pp. 25–41). Spain.

Concrete Institute of Queensland. (2010). Geopolymer concrete. *Concrete IQ*, 3. Retrieved from http://www.ciaq.com.au.

Concrete Society. (2010). Specification for the manufacture curing and testing of GRC products. *Specification*, 13. Retrieved from http://www.grca.co.uk.

Coogee Chemicals. (2012). Sodium aluminate. *Specification*, 1. Retrieved from http://www.coogee.com.au.

Criado, M., Fernandez-Jimenez, A., de la Torre, A. G., Aranda, M. A. G., & Palomo, A. (2007). An XRD study of the effect of the SiO_2/Na_2O ratio on the alkali activation of fly ash. *Cement and Concrete Research, 37*, 671–679.

Cwirzen, A., Habermehl-Cwirzen, K., & Mäkinen, K. (2009). The effect of carbon nano- and microfibers on microcrack formation. *Conference Paper, Concrete Solutions 09* (p. 8). 17–19 Sept., Sydney, Australia

Davidovits, F., & Davidovits, J. (1999). Long lasting Roman cements and concretes. *Géopolymère '99 Proceedings*.

Davidovits, J. (1989). Geopolymers and geopolymeric materials. *Journal of Thermal Analysis, 35*, 429–441.

Davidovits, J. (1991). Geopolymers: Inorganic geopolymeric new materials. *Journal of thermal analysis, 37*, 1633–1656.

Davidovits, J. (2008a). *Geopolymer chemistry and applications* (pp. 398–406). Institut Geopolymere.

Davidovits, J. (2008b). *Geopolymer chemistry and applications* (3rd ed., pp. 149–200). Institut Geopolymere.

Davidovits, J. (2008c). *Geopolymers chemistry and applications* (pp. 392–398). Institut Geopolymere.

Davidovits, J. (2008d). *Geopolymers chemistry and applications* (pp. 488–490). Institut Geopolymere.

Davis, B. (2007). Natural fiber reinforced concrete. *Presentation*, 21. Retrieved from http://people.ce.gatech.edu/~kk92/natfiber.pdf.

de Fazio, P. (2011). Basalt fibre: From earth an ancient material for innovative and modern application. *ENEA report* (No. 8).

de Lacaillarie, J. B. D. E., Bourlon, A., Favier, A., Habert, G., & Roussel, N. (2012). What is geopolymerization? A combined chemical (NMR), structural (SAXS) and rheology study. *Conference Paper*, (p. 16). Monteverita Sunday 12-Opening 15.

de Silva, P., & Sagoe-Crenstil, K. (2008). The Effect of Al_2O_3 and SiO_2 on setting and hardening of Na_2O-Al_2O_3-SiO_2-H_2O geopolymer systems. *Journal of the Australian Ceramic Society, 41*, 39–46.

Debelie, N., Bebruyckere, M., van Nieuwenburg, D., & Deblaere, B. (1997). Attack of concrete floors in pig houses by feed acids: Influence of fly ash addition and cement bound surface layers. *Journal of Agriculutural Engineering Research, 68*, 101–108.

Defazio, C., Arafa, M. D., & Balaguru, P. N. (2006). Functional Geopolymer Composites f. *Technical report*. Center for Advanced Infrastructure and Transportation (CAIT) Civil and Environmental Engineering Rutgers, The State University, Ceram-RU9163 (No. 18).

Demortier, G. (2004). PIXIE, PIGE and NMR study of the masonry of the pyramid of Cheops at Gaza. *Nuclear Instruments and Methods in Physics Research, B226*, 98–109.

Denoël, J. F. (2007). Fire safety and concrete structures. *Technical Review Federation of Belgian Cement Industry*, 90. Retrieved from http://www.febelcem.be.

Dias, D. P., & Thaumaturgo, C. (2005). Fracture toughness of geopolymeric concretes reinforced with basalt fibers. *Cement and Concrete Composites, 27*, 49–54.

Diaz, E. I., Allouche, E. N., & Eklund, S. (2010). Factors affecting the suitability of fly ash as source material for geopolymers. *Fuel, 89*, 992–996.

Dombrowski, K., Buchwald, A., & Weil, M. (2007). The influence of calcium content on the structure and thermal performance of fly ash based geopolymers. *Journal of Materials and Science, 42*, 3033–3043.

Duxon, P., Lukey, G. C., Separovic, F., & van Deventer, J. S. J. (2005). Effect of alkali cations on aluminium incorporation in geopolymeric gels. *Industrial Engineering and Chemical Research, 44*, 832–839.

Duxon, P., Lukey, G. C., & van Deventer, J. S. J. (2006). Thermal conductivity of metakaolin geopolymers used as a first approximation for determining gel interconnectivity. *Industrial and Engineering Chemistry Research, 45*, 7781–7788.

Duxon, P., Lukey, G. C., & van Deventer, J. S. J. (2007). Physical evolution of Na-geopolymer derived from metakaolin up to 1000 C. *Journal of Materials and Science, 42*, 3044–3054.

Duxon, P., Lukey, G. C., & van Deventer, J. S. J. (2007). The thermal evolution of metakaolin geopolymers: Part 2—phase stability and structural development. *Journal of Non-Crystalline Solids, 353*, 2186–2200.

Duxson, P., Fernandez-Jimenez, A., Provis, J. L., Lukey, G. C., Palomo, A., & van Deventer, J. S. J. (2007). Geopolymer technology: The current state of the art. *Journal of Materials and Science, 42*, 2917–2933.

Duxson, P., Lukey, G. C., & van Deventer, J. S. J. (2006a). Nanostructural design of multifunctional geopolymeric materials. *Ceramic Transactions, 175*, 203–214.

Duxson, P., Lukey, G. C., & van Deventer, J. S. J. (2006b). Thermal Evolution of metakaolin geopolymers. Part 1 physical evolution. *Journal of Non-Crystalline Solids, 352*, 5541–5555.

Duxson, P., Provis, J. L., Lukey, G. C., van Deventer, J. S. J., Separovic, F., & Gan, Z. H. (2006). 39 K NMR of free potassium in geopolymers. *Industrial Engineering and Chemical Research, 45*, 9208–9210.

Eden Energy. (2011). Encouraging results on Eden's nano-carbon in concrete. *Press Release*. Retrieved from http://www.edenenergy.com.au.

Eswari, S., Ragunath, P. N., & Suguna, K. (2008). Ductility performance of hybrid fibre reinforced concrete. *American Journal of Applied Sciences, 5*, 1257–1262.

Faraday, M. (1861). In F. A. J. L. James (Ed.), *The chemical history of a candle* (p. 152) Oxford University Press, Sesquicentenary. ISBN: 978-0-19-969491-4.

Fellicetti, R., Meyer, C., & Shimanovich, S. (2001). Basalt fibre reinforced oil well cement slurries. *Conference Paper, Proceedings of the 3rd International Conference on Concrete Under Severe Conditions*, (p. 8) Vancouver.

Fernandez-Jimenez, A., & Palomo, A. (2005). Composition and microstructure of alkali activated fly ash binder: Effect of the activator. *Cement and Concrete Research, 35*, 1984–1992.

Fernandez-Jimenez, A., Pastor, J. Y., Martí'n, A., & Palomo, A. (2010). High-temperature resistance in alkali-activated cement. *Journal of the American Ceramic Society, 93*, 3411–3417.

Fernàndez-Altable, V., & Casanova, I. (2006). Influence of mixing sequence and superplasticiser dosage on the rheological response of cement pastes at different temperatures. *Cement and Concrete Research, 36*, 1222–1230.

Fernández-Jiménez, A., Palomo, A., & Criado, M. (2005). Microstructure development of alkali-activated fly ash cement: a descriptive model. *Cement and Concrete Research, 35*, 1204–1209.

Fernández-Jiménez, A., Palomo, A., Sobrados, I., & Sanz, J. (2006). The role played by the reactive alumina content in the alkaline activation of fly ashes. *Microporous and Mesoporous Materials, 91*, 111–119.

Fernández-Jiménez, A., Palomo, J. G., & Puertas, F. (1999). Alkali-activated slag mortars: Mechanical strength behaviour. *Cement and Concrete Research, 29*, 1313–1321.

Fernandez-Jimenez, A. P. A. (2003). Characterisation of fly ashes. *Potential reactivity as alkaline cements. Fuel, 82*, 2259–2265.

Ferraris, C. F. (1995). Alkali-silica reaction and high performance concrete. *NISTIR5742* . Gaithersburg, MD: Building and Fire Research Laboratory National Institute of Standards and Technology (No. 24).

Fibres Unlimited. (2007). *Test report of basalt fibre reinforced concrete, polypropylene reinforced concrete, polyacrylonitrile reinforced concrete*, (p. 11). Technical Data Sheet. Retrieved from http://www.basalt-fibers.com/sites/default/files/FUBV_Test_Report.pdf.

Fibretech. (2001). *METALX stainless steel fibres offer new improvements in refractory performance, and new opportunities to reduce costs*, (p. 2). Technical Data Sheet. Retrieved from http://www.fibretech.com.

References

Fletcher, R. A., McKenzie, K. J. D., Nicholson, C. L., & Shimada, S. (2005). The composition range of aluminosilicate geopolymers. *Journal of European Ceramic Society, 25*, 1471–1477.

Foerster, S. C., Graule, T., & Gauckler, L. J. (1994). Strength and toughness of reinforced chemically bonded ceramics. *Cement Technology, Ceramic Transactions, American Ceramic Society, 40*, 247–256.

Font, O. M. N., Querol, X., Izquierdo, M. A. E., Diez, S. E. J., Antenucci, D. N. H., & Plana, F. L. A., et al. (2010). X-ray powder diffraction-based method for the determination of the glass content and mineralogy of coal (co)-combustion fly ashes. *Fuel, 89*, 2971–2976.

Forta Corporation. (1999). *Fiber-reinforced shotcrete* (p. 15). Technical Report FORTA Corporation and Technical Data Sheet. Retrieved from http://www.forta-ferro.com.

French, D., & Smitham, J. (2007). Fly ash characteristics and feed coal properties. *CSIRO Energy Technology Research Report, 73*, 55.

Gani, M. S. J. (1997). Concrete. *Materials Forum, 21*, 171–185.

Gastuche, M. C., Toussaint, F., Fripiat, J. J., Touilleaux, R., & van Meersche, M. (1962). Study of intermediate stages in the kaolin-metakaolin transformation. *Clay Minerals, 5*(29), 227–236.

Gay, C., & Sanchez, F. (2010). Performance of carbon nanofiber–cement composites with a high-range water reducer. *Transportation Research Record: Journal of the Transportation Research Board, 2142*, 109–113. Washington, DC: Transportation Research Board of the National Academies.

Effect of Na_2O/Al_2O_3, SiO_2/Al_2O_3 and W/B ratio on setting time and workability of flyash based geopolymer. *International Journal of Engineering Research and Applications (IJERA), 2*, 2142–2147.

Giancaspro, J., Balaguru, P. N., & Lyon, R. E. (2006). Use of inorganic polymer to improve the fire response of balsa sandwich structures. *Journal of Materials in Civil Engineering, 18*, 390–397.

Glatzmaier, G. C., & Ramirez, W. F. (1988). Use of volume averaging for the modeling of thermal properties of porous materials. *Chemical Engineering Science, 43*, 3157–3169.

Gołaszewski, J., & Szwabowski, J. (2004). Influence of superplasticizers on rheological behaviour of fresh cement mortars. *Cement and Concrete Research, 34*, 235–248.

Granizo, M. L., Blanco Varela, M. T., & Martinez-Ramirez, S. (2007). Alkali activation of metakaolins: Parameters affecting mechanical, structural and microstructural properties. *Journal of Materials Science, 42*, 2934–2943.

Greenhalgh, J. (2003, October). Segmental linings—the future is steel fibre reinforcement. *Concrete Magazine*, 19–20.

Guan, S. W. (2003). 100% solids polyurethane and polyurea coatings technology. *Coatings World*, 49–58.

Guerrieri, M., & Sanjayan, J. G. (2010). Behavior of combined fly ash/slag-based geopolymers when exposed to high temperatures. *Fire and Materials, 34*, 163–175.

Hajimohammadi, A., Provis, J. L., & van Deventer, J. S. J. (2008). One-part geopolymer mixes from geothermal silica and sodium aluminate. *Industrial and Engineering Chemistry Research, 47*, 9396–9405.

Hameed, R., Turatsinze, A., Duprat, F., & Sellier, A. (2009). Metallic fibre reinforced concrete: Effect of fiber aspect ratio on the flexural properties. *ARPN Journal of Engineering and Applied Sciences, 4*, 67–72.

Hansen, A. S. (1994). Reinforcing fibres and a method of producing the same. *US Patent5330827*, (p. 11). 17 July 1994.

Hardjito, D. & Rangan, B. V. (2005). Development and properties of low calcium fly ash based geopolymer concrete. *Curtin University Research report GC-1*.

He, X. Y., Sun, W., Gan, X. C., Su, Y., & Yu, M. (2005). Influence of wet-milling on physicochemical properties and strength of fly ash. 6. *Journal of Wuhan University of Technology, 12*.

Healy, J. J., de Groot, J. J., & Kestin, J. (1976). The theory of the transient hot-wire method for measuring thermal conductivity. *Physica, 82C*(1976), 392–408.

Heidrich, C., Ward, C. R., & Gurba, L., (Eds.) (2007). Cooperative research centre for coal in sustainable development. *Coal Combustion Products Handbook*. Callaghan, N.S.W., Australia: Cooperative Research Centre for Coal in Sustainable Development.

Henry, M., Jolivet, J. & Livage, J. (1992). Aqueous chemistry of metal cations: Hydrolysis, condensation and complexation. In R. Reisfeld, & C. K. JØRGENSEN (Eds.), *Chemistry, spectroscopy and applications of sol-gel glasses*. Berlin, Heidelberg: Springer.

Heo, Y. S., Sanjayan, J. G., Han, C. G., & Han, M. C. (2012). Relationship between interaggregate spacing and optimum fibre length for spalling protection of concrete in fire. *Cement and Concrete Research, 42*, 549–557.

Hercules Fibres. (2011). *D5 product overview*. Technical Data Sheet. Retrieved from http://www.herculesfibers.com.

Hertz, K. D. (2003). Limits of spalling of fire-exposed concrete. *Fire Safety Journal, 38*, 103–116.

Higgins, L. (2008). Hazmat articles chemistry of fire. *Information Sheet, Fire and Emergency Services Authority of Western Australia (FESA)*, 2.

Honeywell. (2010). *Spectra® fiber 900 high-strength, light-weight polyethylene fiber*, (p. 2). Technical Data Sheet. Retrieved from http://www.honeywell.com/spectra.

Hu, S. G., Wu, J., & Wang, F. Z. (2010). Preparation technology of high-strength and high-toughness lightweight aggregate concrete. 7. *Proceedings of the 7th International Symposium on Cement & Concrete (ISCC 2010)* May 9–12 Jinan, China.

Ikai, S., Reichert, J. R., Vasconcellos, A. R., & Zampieri, V. A. (2006). Asbestos free technology with new high tenacity PP-polypropylene fibres in air cured Hatschek process. *Conference Paper, 10th International Inorganic Bonded Fibre Composites Conference* (p. 16). Sao Paolo, Brazil.

Institution of Engineers Australia. (1989). *Fire engineering for building structures and safety*, Report. Retrieved from www.engineersaustralia.org.au.

International Organisation for Standardisation. (2010). ISO 11925-2:2010 Reaction to fire tests—Ignitability of products subjected to direct impingement of flame. Part 2: single source flame.

Jamieson, E. J., van Riessen, A., Keally, C., & Hart, R. D. (2012). Development of bayer geopolymer paste and use as concrete. *Proceedings of the Ninth International Alumina Quality Workshop*.

Jansson, A. (2008). Fibres in reinforced concrete structures—analysis, experiments and design (Thesis, Göteborg, Sweden: Department of Civil and Environmental Engineering Division of Structural Engineering, Chalmers University of Technology, 2008). 66.

Jiang, S., Kim, B.-G., & Aıtcin, P.-C. (1999). Importance of adequate soluble alkali content to ensure cement/superplasticizer compatibility. *Cement and Concrete Research, 29*, 71–78.

Jonker, A., Mccrindle, R. I., & van der Merwe, M. J. (2009). Insulating refractory materials from inorganic waste resources. *The Refractories Engineer*, 14–18.

Jung, T. H., & Subramanian, R. V. (1994). Alkali resistance enhancement of basalt fibres by hydrated zirconia films formed by the sol-gel process. *Journal of Material Research, 9*, 1006–1013.

Kalifa, P., Chéné, G., & Gallé, C. (2001). High-temperature behaviour of HPC with polypropylene fibres: From spalling to microstructure. *Cement and Concrete Research, 31*, 1487–1499.

Kamenny Vek. (2012). *Basfiber for construction*, (p. 4). Technical Data Sheet. Retrieved from www.basfiber.com.

Kamseu, E., Ceron, B., Tobias, H. E. L., Bignozzi, M. C., Muscio, A., & Libbra, A. (2012). Insulating behavior of metakaolin-based geopolymer materials assess with heat flux meter and laser flash techniques. *Journal of Thermal Analysis and Calorimetry, 108*, 1189–1199.

Kamseu, E., Nait-Ali, B., Bignozzi, M. C., Leonelli, C., Rossignol, S., & Smith, D. S. (2012). Bulk composition and microstructure dependence of effective thermal conductivity of porous inorganic polymer cements. *Journal of the European Ceramic Society, 32*, 1593–1603.

Kamseu, E., Rizzuti, A., Leonelli, C., & Perera, D. (2010). Enhanced thermal stability in K_2O-metakaolin-based geopolymer concretes by Al_2O_3 and SiO_2 fillers addition. *Journal of Materials and Science, 45*, 1715–1724.

Kantro, D. L. (1980). Influence of water-reducing admixtures on properties of cement paste, a miniature slump test. *Cement, Concrete and Aggregates, 2*, 95–102.

Karbhari, V. M. (1998). Short carbon fiber reinforced concrete. *WTEC monograph, use of composite materials in civil infrastructure in Japan* (p. 211). Chapter 2.

References

Katzer, J. (2006). Steel fibers and steel fiber reinforced concrete in civil engineering. *The Pacific Journal of Science and Technology, 7*, 53–58.

Kendall, A., Keoleian, G. A., & Lepech, M. D. (2008). Materials design for sustainability through life cycle modeling of engineered cementitious composites. *Materials and Structures, 41*, 1117–1131.

Khale, D. C. R. (2007). Mechanism of geopolymerization and factors influencing its development: a review. *Journal of Materials and Science, 42*, 729–746.

Khan, M. I. (2002). Factors affecting the thermal properties of concrete and applicability of its prediction models. *Building and Environment, 37*, 607–614.

Kim, J.-H. J., Mook Lim, Y., Won, J. P., & Park, H. G. (2010). Fire resistant behavior of newly developed bottom-ash-based cementitious coating applied concrete tunnel lining under RABT fire loading. *Construction and Building Materials, 24*, 1984–1994.

Kinrade, S. D., & Pole, D. L. (1992). Effect of alkali-metal cations on the chemistry of aqueous silicate solutions. *Inorganic Chemistry, 31*, 4558–4563.

Kirk-Othmer. (1991). Sodium aluminate production. *Encyclopedia of chemical technology* (4th ed., Vol. 2). New York: Wiley.

Klemens, P. G., & Gell, M. (1998). Thermal conductivity of thermal barrier coatings. *Materials Science and Engineering, A245*, 143–149.

Koehler, E. P. (2009). Test methods for workability and rheology of fresh concrete. *ACI Fall Convention, 35*. Retrieved from http://www.concrete.org.

Koehler, E. P., & Fowler, D. W. (2003). *Summary of concrete workability test methods* (p. 93). ICAR 105-1 International Center for Aggregates Research, The University of Texas at Austin.

Kong, D., Sanjayan, J. G., & Sagoe-Crentsil, K. (2007). Comparative performance of geopolymers made with metakaolin and fly ash after exposure to elevated temperatures. *Cement and Concrete Research, 37*, 1583–1589.

Kong, D., Sanjayan, J. G. & Sagoe Cretsil, K. (2005). Damage due to elevated temperatures in metakaolinite based geopolymer pastes. *Geopolymer Cements and Concrete Conference 2005* (p. 11). Perth.

Kong, D. L. Y., & Sanjayan, J. G. (2008). Damage behavior of geopolymer composites exposed to elevated temperatures. *Cement and Concrete Composites, 30*, 986–991.

Kong, D. L. Y., & Sanjayan, J. G. (2010). Effect of elevated temperatures on geopolymer paste, mortar and concrete. *Cement and Concrete Research, 40*, 334–339.

Konsta-Gdoutos, M. S., Metaxa, Z. S., & Shah, S. P. (2010). Highly dispersed carbon nanotube reinforced cement based materials. *Cement and Concrete Research, 40*, 1052–1059.

Kriven, W. M., Bell, J., & Gordon, M. (2008). Geopolymer refractories for the glass manufacturing industry. *Conference Paper, 64th Conference on Problems with Glass* (p. 8).

Krivenko, P. (2005). Development of alkaline cements supported by theory and practice. *Conference Paper, Proceedings of the International Workshop on Geopolymers and Geopolymer Concrete*.

Kuder, K. G., & Shah, S. P. (2010). Processing of high-performance fiber-reinforced cement-based composites. *Construction and Building Materials, 24*, 181–186.

Kumar, R., Kumar, S., & Mehrotra, S. P. (2007). Towards sustainable solutions for fly ash through mechanical activation. *Resources, Conservation and Recycling, 52*, 157–179.

Kumar, S., Kolay, P., Malla, S., & Mishra, S. (2012). Effect of multiwalled carbon nanotubes on mechanical strength of cement paste. *Journal of Materials in Civil Engineering*, 84–91.

Kumar, S., & Kumar, R. (2010). Tailoring geopolymer properties through mechanical activation of fly ash. *Conference Paper. Second International Conference on Sustainable Construction Materials and Technologies* (p. 8).

Kuraray. (2007). Kuraray, structural fibres for concrete reinforcement. Retrieved from http://www.kuraray-am.com.

Kutchko, B. K. A. G. (2006). Fly ash characterisation by SEM_EDS. *Fuel, 85*, 2537–2544.

Lawson, J. R. (2009). A history of fire testing: past, present, and future. *Journal of ASTM International, 6*, 39.

Lemougna, P. N., Mackenzie, K. J. D., & Chinje Melo, U. F. (2011). Synthesis and Thermal Properties of inorganic polymers (geopolymers for structural and refractory applications from volcanic ash. *Ceramics International, 37*, 3011–3018.

Lepech, M. D., Li, V. C., Robertson, R. E., & Keoleian, G. A. (2008). Design of green engineered cementitious composites for improved sustainability. *ACI Materials Journal, 105*, 567–575.

Li, V. C. (1998). Engineered cementitious composites for structural applications*. *ASCE Journal of Materials in Civil Engineering, 10*, 66–69.

Li, V. C., & Maalej, M. (1996). Toughening in cement based composites. Part II: fiber reinforced cementitious composites. *Cement and Concrete Composites, 18*, 239–249.

Li, Z. (2011). *Advanced concrete technology*. Weinheim, NJ: John Wiley & Sons.

Liefke, E. (1999). Industrial Applications of foamed Inorganic polymers. *Conference Paper Geopolymere 99 Proceedings* (pp. 189–199).

Lin, T., Jia, D., Wang, M., He, P., & Liang, D. (2009a). Effects of fibre content on mechanical properties and fracture behaviour of short carbon fibre reinforced geopolymer matrix composites. *Bulletin of Materials Science, 32*, 77–81.

Lin, T. S., Jia, D. C., He, P. G., & Wang, M. R. (2009b). Thermomechanical and microstructural characterisation of geopolymers with alpha alumina particulate filler. *International Journal of Thermophys, 30*, 1568–1577.

Lipatov, Y. A., Gutnikov, S. I., Manylov, M. S., & Lazoryak, B. I. (2012). Effect of ZrO_2 on the alkali resistance and mechanical properties of basalt fibers. *Inorganic materials* (Vol. 48, pp. 751–756). Pleiades Publishing, Ltd. In Y. V. Lipatov, S. I. Gutnikov, M. S. Manylov, B. I. Lazoryak (Eds.), *Original Russian text* (Vol. 48, No. 7, pp. 858–864). Neorganicheskie Materialy.

Liu, J., Sun, W., Miao, C., Zhang, Q., & Liu, J. (2008). Influence of Superplasticisers and mineral admixtures on the workability of mortar at low water-binder ratio. *Proceedings Second International Conference on Microstructural Related Durability of Cementitious Composites*. 11–13 April 2012, Amsterdam, The Netherlands.

Louisiana Technical University. (2012). Trenchless Technical Centre. Retrieved from http://www.ttc.latech.edu.

Lyon, R. E. (1996). Fire response of geopolymer structural composites. *Technical note DOT/FAA/AR-TN95/22* (No. 15).

Lyon, R. E., Balaguru, P. N., Foden, M., Sorathia, U., Davidovits, J., & Davidovits, D. (1997). Fire resistant aluminosilicate composites. *Fire and Materials, 21*, 67–73.

Lyon, R. E. F., Balaguru, A. J., Davidovits, P., Davidovits, J., Rodriguez, M. (1999). Properties of geopolymer matrix in carbon fibre composites. *Geopolymer 99 Proceedings*.

Ma, S., Zheng, S., Xu, H., & Zhang, Y. (2007). Spectra of sodium aluminate solutions. *Transactions of Nonferrous Metals Society of China, 17*, 853–857.

Mackenzie, K. J. D. (2011). Inorganic polymers for environmental protection applications. *IOP Conference Series: Materials Science and Engineering* (Vol. 18, p. 7).

Maclaren, D. C., & White, M. A. (2003). Cement: its chemistry and properties. *Journal of Chemical Education, 80*, 623–635.

Maitland, C. F., Buckley, C. E., O'Connor, B. H., & Hart, R. D. (2011). Characterisation of porestructure of metakaolin derived geopolymers by neutron scattering and electron microscopy. *Journal of Applied Crystallography, 44*, 697–707.

Manzur, T., & Yazdani, N. (2010). Strength enhancement of cement mortar with carbon nanotubes early results and potential. *Transportation Research Record: Journal of the Transportation Research Board, 2142*, 102–108. Washington, DC: Transportation Research Board of the National Academies.

Marciano, S., Mugnier, N., Clerin, P., Cristol, B., & Moulin, P. (2006). Nanofiltration of Bayer process solutions. *Journal of Membrane Science, 281*, 260–267.

Mazaheripour, H., Ghanbarpour, S., Mirmoradi, S. H., & Hosseinpour, I. (2011). The effect of polypropylene fibers on the properties of fresh and hardened lightweight self-compacting concrete. *Construction and Building Materials, 25*, 351–358.

References

Mccormick, L. (2007). Metakaolin. *Materials Science of Concrete*. Retrieved from www.people.ce.gatech.edu/~kk92/mkgrad.pdf.

Medri, V., Fabbri, S., Ruffini, A., Dedecek, J., & Vaccari, A. (2011). SiC-based refractory paints prepared with alkali aluminosilicate binders. *Journal of the European Ceramic Society, 31*, 2155–2165.

Mehta, P. K., & Meryman, H. (2009). Tools for reducing carbon emissions due to cement consumption. *Structure Magazine*, January 11–15.

Memon, F. A., Nuruddin, M. F., Demie, S., & Shafiq, N. (2012). Development of fly ash-based self-compacting geopolymer concrete. Retieved from http://eprints.utp.edu.my.

Metaxa, Z. S., Konsta-Gdouts, M. S., & Shah, S. P. (2010). Carbon nanofibre reinforced cement based materials. *Journal of the Transportation Research Board*, 114–118.

Minifibers. (2006). *The technology of fybrel™ synthetic pulp in fiber cement*, (p. 5). Technical Data Sheet. Retrieved from www.minifibers.com.

Minifibers. (2010). *Polyethylene fibers low-melt LLDPE*, (p. 1). Technical Data Sheet. Retrieved from www.minifibers.com.

Montes, C., & Allouche, E. (2008). Applications of geopolymer concrete in the rehabilitation of waste water conveyance systems in extreme environments. *Conference Paper, 32nd International Conference and Exposition on Advanced Ceramic and Composites Daytona* (p. 24).

Moolenaar, R. J., Evans, J. C., & McKeever, L. D. (1970). The structure of the aluminate ion in solutions at high pH. *Journal of Physical Chemistry, 74*, 3629–3636.

Morrison, I. J. (2008). GRC standards and testing. *Concrete*, June 12–13.

Moura D., Vasconelos, E., Pacheco-Torgal, F., & Ding, Y. (2011). *Concrete repair with geopolymeric mortars. Influence of mortars composition on their workability and mechanical strength* (p. 6). Retrieved from http://repositorium.sdum.uminho.pt.

Naaman, N. A. (2003). Engineered Steel Fibres with optimal properties for reinforcement of cement composites. *Journal of Advanced Concrete Technology, 1*, 241–252.

Naaman, N. A. (2008). High performance fiber reinforced cement composites. *High performance construction materials science and applications* (pp. 91–153), Chap. 3. Singapore: World Scientific Publishing.

Naik, T. R. (2002). Greener concrete using recycled materials. *Concrete International*. July.

Nasvi, M. C. M., Ranjith, P. G., & Sanjayan, J. (2012). The permeability of geopolymer at downhole stress conditions: Application for carbon dioxide sequestration wells. *Applied Energy, 7*, 62–70.

Nazari, A., Riahi, S., Riahi, S., Shamekhi, S. F., & Khademno, A. (2010). Mechanical properties of cement mortar with Al_2O_3 nanoparticles. *Journal of American Science, 6*, 94–97.

Neville, A. M. (1995). *Properties of concrete* (pp. 148–149, 394–396, 513, 537–538), Harlow: Longman Group.

Nguyen Thang, X., Kroisova, D., Louda, P., & Bortnovsky, O. (2010). Microstructure and flexural properties of geopolymer matrix-fibre reinforced composite with additives of alumina (Al_2O_3) nanofibres. *TEXSCI 2010, 7th International Conference Czech Republic* (p. 8).

Nianyi, C., & Honglin, L. (1994). Structure and relative stability of the aluminate anion studied by quantum chemical methods. *Journal of Molecular Structure (Theochem), 305*, 283–286.

Nicholson, C., Fletcher, R., Miller, N., Stirling, C., Morris, J., Hodges, S., Mackenzie, K., & Schumucker, M. (2005). Building innovation through geopolymer technology. *Chemistry in New Zealand*, September.

Nili, M., & Afroughsabet, V. (2010). The effects of silica fume and polypropylene fibers on the impact resistance and mechanical properties of concrete. *Construction and Building Materials, 24*, 927–933.

Nipponelectricglass. (2000). *Control of plastic shrinkage cracking of concrete with ARG chopped strands*, (p. 7). Technical Data Sheet. Retrieved from www.negamerica.com.

Nipponelectricglass. (2007). *High zirconia alkali resistant glass fibre*, (p. 12). Technical Data Sheet. Retrieved from www.negamerica.com.

Nugteren, H. W., Butselaar-Orthlieb, V. C. L., & Izquierdo, M. (2009). High strength geopolymers from coal combustion fly ash. *Global Nest Journal, 11*, 155–161.

Nugteren, H. W., Ogundiran, M. B., Witkamp, G.-J., & Kreutzer, M. T. (2011). Coal fly ash activated by waste sodium aluminate solutions as an immobiliser for toxic waste. *World of Coal Ash Conference* (pp. 1–10). Denver, USA.

Nycon AR-DM. Technical Data Sheet (p. 2). Retrieved from www.nycon.com.

Nycon. (2009). Nycon G (p. 2). Technical Data Sheet. Retrieved from www.nycon.com.

Nycon. (2011). *Nycon-PVA RF4000 PVA (polyvinyl alcohol), large denier macro, superior bond*, (p. 2). Technical Data Sheet. Retrieved from www.nycon.com.

Nycon. (2012). Nycon-PVA macro, micro fiber blend toughens slabs. http://www.concreteproducts.com.

O'Connor, S. J., Mackenzie, K. J. D., Smith, M. E., & Hanna, J. E. (2010). Ion exchange in charge balancing sites of aluminosilicate inorganic polymers. *Journal of Materials Chemistry, 20*, 10234–10240.

Ohio Coal Development Office. (2004). Fly ash enhanced carbon nanofiber-reinforced high strength concrete. *Project CDO/D-99-14 report* (No. 7).

Okada, K., Imase, A., Isobe, T., & Nakajima, A. (2011). Capillary rise properties of porous geopolymers prepared by an extrusion method using polylactic acid (PLA) fibers as the pore formers. *Journal of the European Ceramic Society, 31*, 461–467.

Orica Chemicals. (2011). *Sodium aluminate s grade*. Product Data Sheet. Retrieved from www.orica.com.

Oualit, M., Jauberthie, R., Rendell, F., Melinge, Y., & Abadlia, M. T. (2012). External corrosion to concrete sewers: a case study. *Urban Water Journal*, 1–6.

Pacheco-Torgal, F., Castro-Gomes, J., & Jalali, S. (2008a). Alkali-activated binders: A review: Part 1. Historical background, terminology, reaction mechanisms and hydration products. *Construction and Building Materials, 22*, 1305–1314.

Pacheco-torgal, F., Castro gomes, J. P., & Jalali, S. (2008b). Durability of historic mortars. *Historic Mortars Conference* (pp. 1–7).

Palacios, M., & Puertas, F. (2005). Effect of superplasticizer and shrinkage-reducing admixtures on alkali-activated slag pastes and mortars. *Cement and Concrete Research, 35*, 1358–1367.

Palomo, A., & Fernandez-Jimenez, A. (2011). Alkali activation procedure for transforming fly ash into new materials. Part 1: Applications. *World of Coal Ash(WOCA) Conference* (p. 14).

Palomo, A., Grutzeck, M. W., & Blanco, M. T. (1999). Alkali activated fly ashes: A cement for the future. *Cement and Concrete Research, 29*, 1323–1329.

Pan, Z., & Sanjayan, J. G. (2012). Factors influencing softening temperature and hot-strength of geopolymers. *Cement and Concrete Composites, 34*, 261–264.

Pan, Z., Sanjayan, J. G., & Rangan, B. V. (2009). An investigation of the mechanisms for strength gain or loss of geopolymer mortar after exposure to elevated temperature. *Journal of Materials and Science, 44*, 1873–1880.

Panneer Selvam, R., & Hale, M. (2011). Evaluation of high temperature concrete for thermal energy storage for solar power generation. *US DOE CSP Programme* (p. 29).

Papworth, F. (2000). *Affect of synthetic fibres and silica fume on explosive spalling of HPC exposed to fire* (p. 6). Singapore: South East Asia Construction.

Peng, G. F., Yang, W.-W., Zhao, J., Liu, Y.-F., Bian, S.-H., & Zhao, L.-H. (2006). Explosive spalling and residual mechanical properties of fiber-toughened high-performance concrete subjected to high temperatures. *Cement and Concrete Research, 36*, 723–727.

Perry, B. (2006). Synthetic macrofibres storm to the front of coastal defence innovation. *Concrete Magazine*, November, 72–73.

Peters, S. J., Rushing, T. S., Landis, E. N., & Cummins, T. K. (2010). Nanocellulose and microcellulose fibers for concrete. *Journal of the Transportation Research Board, 2142*, 25–28. Washington: Transportation Research Board of the National Academies.

Phair, J. W., & van Deventer, J. S. J. (2001). Effect of silicate activator pH on the leaching and material characteristics of waste-based inorganic polymers. *Minerals Engineering, 14*, 289.

References

Phair, J. W., & van Deventer, J. S. J. (2002). Characterisation of fly ash based geopolymeric binders activated with sodium aluminate. *Industrial and Engineering Chemistry Research, 41*.

Phair, J. W., van Deventer, J. S. J., & Smith, J. D. (2004). Effect of Al source and alkali activation on Pb and Cu immobilisation in fly-ash based "geopolymers". *Applied Geochemistry, 19*, 423–434.

Portland Cement Association. (2012). *What is ettringite and does it or the sulphate in cement contribute to expansion and disintegration of portland cement concrete?* (p. 1). Retrieved from http://www.cement.org/tech/faq_DEF.asp.

PQ Corporation. (2005). Fundamentals of silicate chemistry. *Brochure*. Retrieved from www.pqcorp.com.

PQ Europe. (2004). Sodium and potassium silicates, versatile compounds for your applications. *Brochure*. Retrieved from www.pqcorp.com.

Provis, J. L. (2009). Activating solution chemistry for geopolymers. In J. L. Provis & J. S. J. van Deventer (Eds.), Chapter 4. *Geopolymers structure, processing and industrial applications* (pp. 50–71). Woodhead Publishing Ltd.

Provis, J. L. (2010). Fire resistance of geopolymer concretes. *Project report AOARD-084096*. University of Melbourne (No. 8).

Provis, J. L. (2014). Geopolymer schematic. *Alkali activated materials: State of the art report* (pp. 1–9). Dordrecht: Rilem TC224-AAM, Springer/RILEM.

Provis, J. L., Harrex, R. M., Bernal, S. A., Duxson, P., & van Deventer, J. S. J. (2012). Dilatometry of geopolymers as a means of selecting desirable fly ash sources. *Journal of Non-Crystalline Solids, 358*, 1930–1937.

Puertas, F., Amat, T., Fernández-Jiménez, A., & Vázquez, T. (2003). Mechanical and durable behaviour of alkaline cement mortars reinforced with polypropylene fibres. *Cement and Concrete Research, 33*, 2031–2036.

Puertas, F., Martinez-Ramirez, S., Alonso, S., & Vasquez, T. (2000). Alkali activated fly ash/slag cement strength behavior and hydration products. *Cement and Concrete Research, 30*, 1625–1632.

Purdon, A. O. (1940). The action of alkalis on blast furnace slag. *Journal of the Society of the Chemical Industry, 59*, 191–202.

Rahier, H., Simons, W., van Mele, B., & Biesemens, M. (1997). Low temperature synthesized aluminosilicate glasses. Part 3 influence of the composition of the silicate solution on production, structure and properties. *Journal of Material Science, 32*, 2237–2247.

Rahier, H., van Mele, B., Biesemans, M., Wastiels, J., & Wu, X. (1996). Low-temperature synthesized aluminosilicate glasses. Part I low-temperature reaction stoichiometry and structure of a model compound. *Journal of Material Science, 31*, 71–79.

Rahier, H., Wastiels, J., Biesemans, M., Willlem, R., van Assche, G., & van Mele, B. (2007). Reaction mechanism, kinetics and high temperature transformations of geopolymers. *Journal of Material Science, 42*, 2982–2996.

Rayzman, V., Filipovich, I., Nisse, L., & Vlasenko, Y. (1998). Sodium aluminate from alumina-bearing intermediates and wastes. *JOM*. November 1998.

Reid, M. (2011). Window or door frame. *UK Patent Application GB 2478535 A CSENG Ventures* (p. 13).

Reoforcetech (2012a). *Basalt fibre reinforcement technology*, (p. 2). Technical Data Sheet. Retrieved from www.reforcetech.com.

Reoforcetech. (2012b). *BFRP minibars*, (p. 2). Technical Data Sheet. Retrieved from www.reforcetech.com.

Rickard, W. D. A., Temuujin, J. & van Riessen, A. (2012). Thermal analysis of geopolymer pastes synthesised from five fly ashes of variable composition. *Journal of Non-Crystalline Solids, 10*.

Rickard, W. D. A., van Riessen, A., & Walls, P. (2010). Thermal character of geopolymers synthesized from class F fly ash containing high concentrations of Iron and alpha-quartz. *International Journal of Applied Ceramic Technology, 7*, 81–88.

Rickard, W. D. A., Vickers, L., & van Riessen, A. (2013). Performance of fibre reinforced, low density metakaolin geopolymers under simulated fire conditions. *Applied Clay Science, 73*, 71–77.

Rickard, W. D. A., Williams, R., Temuujin, J., & van Riessen, A. (2011). Assessing thesuitability of three Australian fly ashes as an aluminosilicate source for geopolymers in high temperature applications. *Materials Science and Engineering: A, 528*, 3390–3397.

Rill, E., Lowry, D. R., & Kriven, W. M. (2010). Properties of basalt fiber reinforced geopolymer composites. *Strategic materials and computational design*. Wiley.

Rixom and Mailvaganam. (1999). *Chemical admixtures for concrete*. Spon Press, ISBN: 9780419126300.

Rodriquez, R., Howser, R., & Mo, Y. L. (2010). Structural behavior of self-consolidating carbon nanofiber concrete. *University of Houston report* (No. 32).

Rollett, A. D. (2007). L8: thermal properties. Carnegie Mellon University lecture power point.

Romualdi, J. P., & Batson, G. B. (1963). Behaviour of reinforced concrete beams with closely spaced reinforcement. *Journal of the American Concrete Institute, 60*, 775–790.

Romualdi, J. P., & Mandel, J. A. (1964). Tensile strength of concrete affected by uniformly distributed and closely spaced short lengths of wire reinforcement. *Journal of the American Concrete Institute, 61*, 657–672.

Ross, A. (2009). Steel fibre reinforced concrete (SFRC)—quality, performance and specification. *Conference Paper New Zealand Concrete Conference 09* (p. 7).

Ross, A. (2012). Steel fibre reinforced concrete (SFRC) combined with conventional reinforcing. *Concrete Australia*, 1.

Rostami, H., & Brendley, W. (2003). Alkali ash material: A novel fly ash cement. *Environmental Science and Technology, 37*, 3454–3457.

Rotstein, C. (2011). New developments in the research of strain herdening concrete. *Conference Paper, 2011 Concrete Institute Conference,* Perth.

Rowles, M., & O'Connor, B. H. (2003). Chemical optimisation of the compressive strength of aluminosilicate geopolymers synthesised by sodium silicate activation of metakaolinite. *Journal of Materials Chemistry, 13*, 1161–1165.

Roy, D. M. (1999). Alkali activated cements, opportunities and challenges. *Cement and Concrete Research, 29*, 249–254.

Scheffler, C., Förster, T., Mäder, E., Heinrich, G., Hempel, S., & Mechtcherine, V. (2009). Aging of alkali-resistant glass and basalt fibers in alkaline solutions: Evaluation of the failure stress by Weibull distribution function. *Journal of Non-Crystalline Solids, 355*, 2588–2595.

Scrivener, K. (2011). Straight talk with Karen Scrivener on cements, CO_2 and sustainable development. *American Ceramic Society Bulletin, 91*, 47–50.

Scrivener, K. L., Crumbie, A. K., & Laugesen, P. (2004). The interfacial transition zone (ITZ) between cement paste and aggregate in concrete. *Interface science* (Vol. 12, pp. 411–421) Kluwer.

Serway, A. S. (1992). *Physics for scientists and engineers with modern physics* (3rd ed., pp. 526–559). Chap. 20. Saunders HBJ Florida.

Shi, C., & Mo, Y. L. (2008). *High performance construction materials science and applications*. Singapore: World Scientific Publishing.

Silva, F. J., Mathias, A. F., & Thaumaturgo, C. (1999). Evaluation of the fracture toughness in poly(sialate-siloxo) composite matrix. *Conference Paper, Geopolymere 99* (pp. 97–106).

Silva, F. J., & Thaumaturgo, C. (2003). Fibre reinforcement and fracture response in geopolymeric materials. *Fatigue and Fracture of Engineering Materials and Structures*, 26.

Silverstrim, T., Rostami, H., Larralde, J., & Samadi, A. (1997). Fly ash cementitious material and a method of making a product. *US Patent 5601643*, p. 10.

Sipos, P. (2009). The structure of Al(III) in strongly alkaline aluminate solutions—a review. *Journal of Molecular Liquids, 146*, 1–14.

Skvara, F., Svoboda, P., Dolezal, J., Kopecky, L., Pawlasova, S., Myskova, L., Lucuk, M., Dvoracek, K., Beksa, M., & Sulc, R. (2006). Concrete based on fly ash geopolymer. *10th East Asia-Pacific Conference on Structural Engineering and Construction* (pp. 407–412).

Sofi, M., van Deventer, J. S. J., Mendis, P. A., & Lukey, G. C. (2007). Engineering Properties of Inorganic polymer Concretes (IPCs). *Cement and Concrete Research, 37*, 251–257.

Southern Ionics. (2006). Sodium aluminate, 38%. *Product Bulletin*. Retrieved from www.southernionics.com.

Standards Australia. (2005). AS1530.4-2005 methods for fire tests on building materials, components and structures—fire resistance test of elements of construction.

Stavinoha, R. (1991). Protecting concrete from exposure to aggressive chemicals. *Concrete Construction*, July, 3.

Steins, P., Poulesquen, A., Diat, O., & Langmuir, F F. (2012). Structural evolution during geopolymerization from an early age to consolidated material. *Langmuir, 28*, 8502–8510.

Stephens, R. (2009). Coal sludge flood and public response, Roane county Tennessee. *Disaster Response, Environmental Health, Environmental Health Professionals, 7*.

Subaer, & van Riessen, A. (2007). Thermo-mechanical and microstructural characterisation of sodium-poly(sialate-siloxo) (Na-PSS) geopolymers. *Journal of Materials and Science, 42*, 3117–3123.

Swamy, R. N., & Mangat, P. S. (1974). Influence of fiber geometry on the properties of steel fiber reinforced concrete. *Cement and Concrete Research, 4*, 451–465.

Sweetman, B. (2011, October). F35B concrete buster. *Defense Technology International, 36*.

Tarada, F., & King, M. (2009). Structural fire protection of railway tunnels. *Conference Paper. Railway Engineering Conference* (p. 10). Halcrow, UK: University of Westminster, 24–25 June 2009.

Tchakoute Kouamoa, H., Mbeya, J. A., Elimbia, A., Kenne Diffoa, B. B., & Njopwouo, D. (2012). Synthesis of volcanic ash-based geopolymer mortars by fusion method: Effects of adding metakaolin to fused volcanic ash. *Ceramics International, 39*(2), 1613–1621.

Temuujin, J., Minjigma, A., Rickard, W., & van Riessen, A. (2012). Thermal properties of spray-coated geopolymer-type compositions. *Journal of Thermal Analysis and Calorimetry, 107*, 287–292.

Temuujin, J., Minjigmaa, A., Rickard, W., Lee, M., Williams, I., & van Riessen, A. (2009). Preparation of metakaolin based geopolymer coatings on metal substrates as thermal barriers. *Applied Clay Science, 46*, 265–270.

Temuujin, J., Minjigmaab, A., Rickard, W., Lee, M., Williams, I., & van Riessen, A. (2010). Fly ash based geopolymer thin coatings on metal substrates and its thermal evaluation. *Journal of Hazardous Materials, 180*, 748–752.

Temuujin, J., & van Riessen, A. (2009a). Effect of fly ash preliminary calcination on the properties of geopolymer. *Journal of Hazardous materials, 164*, 634–639.

Temuujin, J., van Riessen, A., & Wiilliams, R. (2009b). Influence of calcium compounds on the mechanical properties of fly ash geopolymer pastes. *Journal of Hazardous materials, 167*, 82–88.

Temuujin, J., Williams, R. P., & van Riessen, A. (2009c). Effect of mechanical activation of fly ash on the properties of geopolymer cured at ambient temperature. *Journal of Materials Processing Technology, 209*, 5276–5280.

Temuujin, T., Rickard, W. D. A., Lee, M., & van Riessen, A. (2011). Preparation and thermal properties of fire resistant metakaolin based geopolymer type coatings. *Journal of Non-Crystalline Solids, 357*, 1399–1404.

Thang, X. N., Kroisova, D., Louda, P., & Bortnovsky, O. (2010). Microstructure and flexural strength properties of geopolymer matrix-fibre reinforced composite with additives of alumina (Al_2O_3) nanofibres. *Conference Paper 7th International Conference—TEXSCI 2010* (p. 8). Liberec: Czech Republic, September 6–8.

The Concrete Society. (2007). *Technical report 63*. Guidance for the design of steel-fibre-reinforced concrete.

The Quartz Page. (2012). Quartz, dependence of structure on temperature. Retrieved from http://www.quartzpage.de.

Tonoli, G. H. D., Rodrigues Filho, U. P., Savastano, J. R. H., Bras, J., Belgacem, M. N., & Rocco Lahr, F. A. (2009). Cellulose modified fibres in cement based composites. *Composites: Part A, 40*, 2046–2053.

Toutanji, H., McNeil, S., & Bay, Z. (1998). Chloride permeability and impact resistance of polypropylene fiber reinforced silica fume concrete. *Cement and Concrete Research, 28*, 961–968.

Tran, D. H., Kroisová, D., Louda, P., Bortnovsky, O., & Bezucha, P. (2009). Effect of curing temperature on flexural properties of silica-based geopolymer-carbon reinforced composite. *Journal of Achievements in Materials and Manufacturing Engineering, 37*, 492–497.

Transportation Research Board. (2006). Control of cracking in concrete, state of the art. *Transportation Research Circular, E-C107*, 56.

van Alken, S. (2012). Cracking at the unheated side of a tunnel during the heating and cooling phase of a fire. (M.Sc. thesis, Delft University of Technology, 2012).

van Deventer, J. S. J., Provis, J. L., Duxson, P., & Lukey, G. C. (2007). Reaction mechanisms in the geopolymeric conversion of inorganic waste to useful products. *Journal of Hazardous materials, 139*, 506–513.

van Jaarsveld, J. G. S., & van Deventer, J. S. J. (1999). Effect of the alkali metal activator on the properties of fly ash-based geopolymers. *Industrial and Engineering Chemistry Research, 38*, 3932–3941.

van Jaarsveld, J. G. S., van Deventer, J. S. J., & Lorenzen, L. (1997). The potential use of geopolymeric materials to immobilise toxic metals: Part I. Theory and applications. *Minerals Engineering, 10*, 659.

van Jaarsveld, J. G. S., van Deventer, J. S. J., & Lukey, G. C. (2003). The characterisation of source materials in fly ash based geopolymers. *Material Letters, 57*.

van Oss, H. G. (2012). Cement. *US Geological Survey, Mineral Commodity Summaries*, 1–2.

van Riessen, A. (2010). Geopolymer highlights. *CSRP 2010 Geopolymer Conference* (p. 22).

van Riessen, A., & Chen Tan, N. (2013). Beneficiation of Collie fly ash for synthesis of geopolymer: Part 1—beneficiation. *Fuel, 106*, 569–575.

van Riessen, A., Rickard, W., & Sanjayan, J. (2009). Thermal properties of Geopolymers. In J. L. Provis, J. A. S. van Derventer (Eds.), *Geopolymers: structure, processing, properties and industrial applications* (pp. 315–342). Woodhead Publishing Ltd.

van Zijl, G. P. A. G., & Wittmann, F. H. (2010). On durability of SHCC. *Journal of Advanced Concrete Technology, 8*, 261–271.

Vaou, V., Panias, D., & Laboratory Of Metallurg, Y, S. O. M. (2010). Thermal insulating foamy geopolymers from perlite. *Minerals Engineering, 23*, 1146–1151.

Vera-Agullo, J., Chozas-Ligero, V., Portillo-Rico, D., & Garcia-Casas, M. J. (2009). Mortar and concrete reinforced with nanomaterials. *Nanotechnology in Construction, 3*, 383–388.

Verdolotti, L., Iannace, S., Lavorgna, M., & Lamanna, R. (2008). Geopolymerization reaction to consolidate incoherent pozzolanic soil. *Journal of Materials and Science, 43*, 865–873.

Vijai, K., Kumuthaa, R., & Vishnuramb, B. G. (2012). Properties of glass fibre reinforced geopolymer concrete composites. *Asian Journal of Civil Engineering (Building and Housing), 13*, 511–520.

Vitro Minerals. (2010). *Portland cement and pozzolans technical background for the effective use of VCAS pozzolans in Portland cement concrete*. Technical Data Sheet. Retrieved from www.VitroMinerals.com.

Wan, H., Shui, Z., & Lin, Z. (2004). Analysis of geometric characteristics of GGBS particles and their influences on cement properties. *Cement and Concrete Research, 34*, 133–137.

Wang, C., & Niu, D. (2012). Research on the durability of polypropylene fiber concrete under freeze-thaw damage. *Conference Paper, Second International Conference on Microstructural-related Durability of Cementitious Composites* (p. 6). Amsterdam: The Netherlands, 11–13 April 2012.

Wang, M.-R., Jia, D.-C., He, P.-G., & Zhou, Y. (2011). Microstructural and mechanical characterization of fly ash cenosphere/metakaolin-based geopolymeric composites. *Ceramics International, 37*, 1661–1666.

Wang, S. D., & Scrivener, K. L. (1995). Hydration products of alkali activated slag cement. *Cement and Concrete Research, 25*, 561–571.

Wang, X. H., Jacobsen, S., Lee, S. F., He, J. Y., & Liang Zhang, Z. L. (2010). Effect of silica fume, steel fiber and ITZ on the strength and fracture behavior of mortar. *Materials and Structures, 43*, 125–139.

Wang, Y., Li, V. C., & Backer, S. (1991). Tensile failure mechanisms in synthetic fibre-reinforced mortar. *Journal of Materials and Science, 26*, 6565–6575.

Washington Mills. (2008). *Duramul*. Technical Data Sheet. Retieved from www.washingtonmills.com.

References

Weng, L., & Sagoe-Cretsil, K. (2007). Dissolution processes, hydrolysis and condensation reactions during geopolymer synthesis: Part I—low Si/Al ratio systems. *Journal of Materials and Science, 42*, 2997–3006.

Wild, S., Khatib, J. M., & Jones, A. (1996). Relative strength, pozzolanic activity and cement hydration in superplasticised metakaolin concrete. *Cement and Concrete Research, 26*, 1537–1544.

Williams, R. P., & van Riessen, A. (2010). Determination of the reactive component of fly ashes for geopolymer production using XRF and XRD. *Fuel, 89*, 3683–3692.

Wimpenny, D., Angerer, W., Cooper, T., & Bernard, S. (2009). The use of steel and synthetic fibres in concrete under extreme conditions. *Conference Paper Concrete Solutions 09 Paper* (Vol. 4b-5, p. 10).

Won, J. P., Kang, H. B., Lee, S. J., & Kang, J. W. (2012). Eco-friendly fireproof high-strength polymer cementitious composites. *Construction and Building Materials, 30*, 406–412.

Xiao, L. Z., Wei, X. S., & Li, Z. J. (2011). Retarding effect of superplasticizer on the hydration of Portland cement. In C. Leung & K.T. WAN (Eds.), *International RILEM Conference on Advances in Construction Materials Through Science and Engineering*, (Vol. 8). RILEM Publications SARL. ISBN: 978-2-35158-116-2, e-ISBN: 978-2-35158-117-9.

Xu, H., & van Deventer, J. S. J. (2000). The geopolymerisation of alumino silicate materials. *International Journal of Mineral Processing, 59*, 247–266.

Yazdanbakhsh, A., Grasley, Z., Tyson, B., & Abu Al-Rub, R. K. (2010). Distribution of carbon nanofibers and nanotubes in cementitious composites. *Journal of the Transportation Research Board, 2142*, 89–95. Washington, DC: Transportation Research Board of the National Academies.

Yilmaz, V. T., Odabaçoğlu, M., İçbudak, H., & Ölmez, H. (1993). The degradation of cement superplasticizers in a high alkaline solution. *Cement and Concrete Research, 23*, 152–156.

Yip, C. K., Lukey, G. C., Provis, J. L., & van Deventer, J. S. J. (2008). Effect of calcium silicate sources on geopolymerisation. *Cement and Concrete Research, 38*, 554–564.

Yunsheng, Z., Wei, S., & Zongjin, L. (2006). Impact behavior and microstructural characteristics of PVA fiber reinforced fly ash-geopolymer boards prepared by extrusion technique. *Journal of Materials and Science, 41*, 2787–2794.

Yunsheng, Z., Wei, S., Zongjin, L., Xiangming, Z., & Chau Chungkong, E. (2008). Impact properties of geopolymer based extrudates incorporated with fly ash and PVA short fiber. *Construction and Building Materials, 22*, 370–383.

Zhang, S., Li, G. Z., & Yuan, H. Y. (2011). Effect of the chemical treated kevlar fibre on the behaviours of cement. *Advanced Materials Research, 306–307*, 758–761.

Zhang, Z., Yao, X., & Wang, H. (2010). Thermochemistry study of the effect of alkali content on the early geopolymerisation at room temperature. In Shi, C. & Shen, X.(Eds.), *First International Conference on Advances in Chemically-Activated Materials CAM' 2010*, (pp. 171–180). RILEM Publications SARL. Print-ISBN: 978-2-35158-101-8.

Zhang, Z., Yao, X., Zhu, H., Hua, S., & Chen, Y. (2009). Preparation and mechanical properties of polypropylene fiber reinforced calcined kaolin-fly ash based geopolymer. *Journal of Central South University of Technology, 16*, 0049–0052.

Zhang, Z. Y. X., Wang, X., Yao, X., & Zhu, H. (2010). Potential application of geopolymers as protection coatings for marine concrete: II. Microstructure and anticorrosion mechanism. *Applied Clay Science, 49*, 7–12.

Zhao, R., & Sanjayan, J. (2012). Geopolymer and Portland cement concretes in simulated fire. *Magazine of Concrete Research, 63*, 163–173.

Zhu, H. J., Yao, X., & Zhang, Z. H. (2010). Study on non-cement based Alkali Activated material for Oil and Gas well cementing at low to moderate temperatures. *Conference Paper*.

Zivica, V., & Bajza, A. (2001). Acidic attack of cement based materials—a review. Part 1. Principle of acidic attack. *Construction and Building Materials, 15*, 331–340.

Zuhua, Z., Xiao, Y., Huajun, Z., & Yue, C. (2009). Role of water in the synthesis of calcined kaolin-based geopolymer. *Applied Clay Science, 43*, 218–223.

MIX
Papier aus verantwortungsvollen Quellen
Paper from responsible sources
FSC® C105338

If you have any concerns about our products,
you can contact us on
ProductSafety@springernature.com

In case Publisher is established outside the EU,
the EU authorized representative is:
**Springer Nature Customer Service Center GmbH
Europaplatz 3, 69115 Heidelberg, Germany**

Printed by Libri Plureos GmbH
in Hamburg, Germany